Lecture Notes in Networks and Systems

Volume 137

The series "Lecture Notes in Networks and Systems" publishes the latest developments in Networks and Systems—quickly, informally and with high quality. Original research reported in proceedings and post-proceedings represents the core of LNNS.

Volumes published in LNNS embrace all aspects and subfields of, as well as new challenges in, Networks and Systems.

The series contains proceedings and edited volumes in systems and networks, spanning the areas of Cyber-Physical Systems, Autonomous Systems, Sensor Networks, Control Systems, Energy Systems, Automotive Systems, Biological Systems, Vehicular Networking and Connected Vehicles, Aerospace Systems, Automation, Manufacturing, Smart Grids, Nonlinear Systems, Power Systems, Robotics, Social Systems, Economic Systems and other. Of particular value to both the contributors and the readership are the short publication timeframe and the world-wide distribution and exposure which enable both a wide and rapid dissemination of research output.

The series covers the theory, applications, and perspectives on the state of the art and future developments relevant to systems and networks, decision making, control, complex processes and related areas, as embedded in the fields of interdisciplinary and applied sciences, engineering, computer science, physics, economics, social, and life sciences, as well as the paradigms and methodologies behind them.

**** Indexing: The books of this series are submitted to ISI Proceedings, SCOPUS, Google Scholar and Springerlink ****

More information about this series at http://www.springer.com/series/15179

Jyotsna K. Mandal · Somnath Mukhopadhyay ·
Alak Roy
Editors

Applications of Internet of Things

Proceedings of ICCCIOT 2020

Springer

Editors
Jyotsna K. Mandal
Department of Computer
Science and Engineering
University of Kalyani
Nadia, West Bengal, India

Somnath Mukhopadhyay
Department of Computer
Science and Engineering
Assam University
Silchar, Assam, India

Alak Roy
Department of Information Technology
Tripura University
Agartala, Tripura, India

ISSN 2367-3370 ISSN 2367-3389 (electronic)
Lecture Notes in Networks and Systems
ISBN 978-981-15-6197-9 ISBN 978-981-15-6198-6 (eBook)
https://doi.org/10.1007/978-981-15-6198-6

This Springer imprint is published by the registered company Springer Nature Singapore Pte Ltd.
The registered company address is: 152 Beach Road, #21-01/04 Gateway East, Singapore 189721, Singapore

Preface

Tripura University, India, organized the First International Conference on "International Conference on Computer Communication and Internet of Things (ICCCIoT 2020)," during 03–04 February 2020, at the Department of Information Technology. This mega event covered all aspects of communications and Internet of Things (IoT), where scope was not limited to various engineering disciplines such as computer science, electronics and engineering researches but also included researches from allied community like data analytics and network security. The primary goal of "ICCCIoT 2020" was to present the state-of-the-art scientific findings, encourage academic and industrial interaction and to promote collaborative research activities in computer communication, Internet of Things and related fields, involving scientists, engineers, professionals, researchers and students across the globe.

This volume is a collection of high-quality peer-reviewed research papers received across the globe. Based on rigorous peer-review process by the technical programme committee members along with external experts as reviewers (inland as well as abroad), best-quality papers were identified for presentation and publication. The review process was extremely stringent with minimum three reviews for each submission and occasionally up to six reviews. Checking of similarities and overlaps is also done based on the international norms and standards. The organizing committee of ICCCIoT 2020 was constituted with a strong international academic and industrial luminaries and the technical programme committee comprised more than hundred domain experts. The proceedings of the conference is published in Lecture Notes in Networks and Systems, Springer (LNNS). We, in the capacity of the volume editors, convey our sincere gratitude to Springer Nature for providing us the opportunity to publish the proceedings of ICCCIoT in LNNS series.

This conference included distinguished speakers such as Prof. Rajkumar Buyya, Director, Cloud Computing and Distributed Systems (CLOUDS) Lab, The University of Melbourne, Australia; Prof. Bhabani P. Sinha, Former Professor,

Statistical Institute, Kolkata, India; Dr. Sushanta Karmakar, Indian Institute of Technology Guwahati, India; Prof. Sudipta Roy, Assam University, Silchar, India; and Prof. Shikhar Kumar Sarma, Professor, Guwahati University.

Our sincere gratitude to Prof. Biman Kumar Dutta, Hon'ble Member, North East Council, India; Shri. Atanu Saha, Director (S&T), North East Council, India; and Shri. P. L. N. Raju, Director, NESAC, India, for their suggestions regarding various processes during the conference and also funding for the conference. The editors also thank the chairs of the technical sessions of ICCCIoT 2020 for taking the troubles to guide the authors and presenters to enrich their articles for preparing final camera-ready versions.

Special mention of words of appreciation is due to Prof. Mahesh Kumar Singh, Chief Patron of the conference; Prof. Sukanta Banik and Prof. Chandrika Basu Majumder, Honorary Chairs of the conference; Dr. Swanirbhar Majumder, Organizing Chair; and Dr. Alak Roy and Mr. Jayanta Pal, Joint Organizing Secretary of the conference for hosting the conference. Special thanks to Ashish Choudhury, Information Scientist, Tripura University, for his quick response for similarity checking of the papers. It was indeed heartening to note the enthusiasm of all faculty, staff and students of Tripura University to organize the conference in a professional manner. The involvement of faculty coordinators and student volunteers is particularly praiseworthy in this regard. The editors leave no stone unturned to thank technical partners and sponsors for providing all the support and financial assistance.

It is needless to mention the role of the contributors for their active support and participation by submitting their research findings. We take this privilege to thank the authors of all the papers submitted as a result of their hard work. We are further indebted to the technical programme committee members and external reviewers who not only produced excellent reviews but also maintained the high academic standard of the proceedings in short timeframes, in spite of their very busy schedule.

The conference may meet its completeness if it is able to attract elevated participation in its fold. We would like to appreciate the participants of the conference, who have considered the conference a befitting one in spite of all the hardship they had undergone. Last but not least, we would offer cognizance to all the volunteers for their tireless efforts in meeting the deadlines and arranging every minute detail meticulously to ensure that the conference achieves its goal, academic or otherwise and that too unhindered.

Hope this volume will be a useful material to the researchers, practicing engineers and students.

Nadia, India — Jyotsna K. Mandal
Silchar, India — Somnath Mukhopadhyay
Agartala, India — Alak Roy
Editors

Contents

Editors and Contributors

About the Editors

Dr. Jyotsna K. Mandal received his M.Tech. in Computer Science from the University of Calcutta and his Ph.D. from Jadavpur University in the field of Data Compression and Error Correction Techniques. Currently, he is a Professor of Computer Science and Engineering and Director of the IQAC at the University of Kalyani, West Bengal, India. He is a former Dean of Engineering, Technology & Management (2008–2012). He has 33 years of teaching and research experience. He has served as a Professor of Computer Applications, Kalyani Government Engineering College for two years and as an Associate and Assistant Professor at the University of North Bengal for sixteen years. He has been a life member of the Computer Society of India since 1992 and senior member of IEEE. Further, he is a Fellow of the IETE and a member of the AIRCC. He has produced 176 publications in various international journals, has edited thirty-four volumes as a Volume Editor for Science Direct, Springer, CSI, etc., and has successfully executed five Research Projects funded by the AICTE, Ministry of IT Government of West Bengal. In addition, he is a Guest Editor of Microsystem Technology Journal. 23 scholars awarded Ph.D. degree under his supervision and eight are pursuing.

Dr. Somnath Mukhopadhyay is currently an Assistant Professor at the Department of Computer Science and Engineering, Assam University, Silchar, India. He completed his M.Tech. and Ph.D. degrees in Computer Science and Engineering at the University of Kalyani, India, in 2011 and 2015, respectively. He has co-authored one book and has six edited books to his credit. He has published over 30 papers in various international journals and conference proceedings, as well as five chapters in edited volumes. His research interests include digital image processing, computational intelligence, and remote sensing. He is a member of

IEEE and IEEE Computational Intelligence Society, Kolkata Section; life member of the Computer Society of India; and currently the regional student coordinator (RSC) of Region II, Computer Society of India.

Dr. Alak Roy, B.Tech. in Computer Science and Engineering from North Eastern Regional Institute of Science and Technology in 2008, M.Tech. in Information Technology from Tezpur University in 2010, awarded Ph.D. in Computer Science and Engineering from Tezpur University in 2010. Qualified UGC NET and GATE in 2017. Presently, he is working as an Assistant Professor in the Department of Information Technology at Tripura University, India, from May, 2012. He has served as an Assistant Professor, Department of Computer Science & Engineering at the National Institute of Technology Agartala from October 2010 to April 2012. He has nine years of teaching and research experience in Wireless Ad-Hoc and Sensor Networks, Internet of Things, Wireless and Mobile Communication, Underwater Sensor Networks, and Computer Networks. He has supervised more than 26 master dissertations. Dr. Roy has published more than 25 papers in international journals and conference proceedings and organized 2 International conferences and 13 workshops. He serves as a Reviewer of 6 journals and 10 conferences and professional member of IEEE, ACM, IAENG, and IAASSE.

Contributors

Anuradha Banerjee Kalyani Government Engineering College, Kalyani, India

Anwesha Banik Department of Computer Science and Engineering, Tripura Institute of Technology, Narsingarh, Tripura, India

Rita Banik National Institute of Technology Agartala, Agartala, Tripura, India

Somudeep Bhattacharjee Department of Electrical Engineering, Tripura University, Agartala, Tripura, India

Madhurima Bhattacharya Department of Information Technology, Tripura University, Agartala, Tripura, India

Bishanka Brata Bhowmik Department of Electronics and Communication Engineering, Tripura University, Agartala, Tripura, India

Ankur Biswas Department of Computer Science and Engineering, Tripura Institute of Technology, Narsingarh, Tripura, India

M. H. A. Biswas Mathematics Discipline, Khulna University, Khulna, Bangladesh

Sangita Das Biswas Department of Electrical Engineering, Tripura University, Agartala, Tripura, India

Samarjeet Borah Department of Computer Application, SMIT, Sikkim Manipal Institute of Technology, Sikkim, India

Prakash Chauhan Tezpur University, Tezpur, Assam, India

Krishna Das Department of Computer Science and Engineering, Tezpur University, Napaam, Tezpur, Assam, India

Priyanath Das National Institute of Technology Agartala, Agartala, Tripura, India

Purnendu Das Department of Computer Science, Assam University Silchar, Silchar, Assam, India

Srikanta Das Tripura University, Agartala, Tripura, India

Tushar Kanti Das Electrical Engineering, Techno College of Engineering Agartala, Madhuban, Tripura, India

Birajit Debbarma Department of Computer Science and Engineering, Tripura Institute of Technology, Narsingarh, Tripura, India

Papiya Debbarma Tripura University, Agartala, Tripura, India

Monalisha Debnath Department of Computer Science and Engineering, Tripura Institute of Technology, Narsingarh, Tripura, India

Rajesh Debnath Department of Electrical Engineering, Tripura University, Agartala, Tripura, India

Ranadhir Debnath Department of Information Technology, Tripura University, Agartala, Tripura, India

Suman Debnath Department of Electronics and Communication Engineering, Tripura University, Agartala, Tripura, India

Sanjib K. Deka Tezpur University, Tezpur, Assam, India

Mampi Devi Department of Computer Science and Engineering, Tripura University, Agartala, Tripura, India

Amina Elbatoul Dinar Faculty of Sciences and Technology, Mustapha Stambouli University, Mascara, Algeria;
LSTE Laboratory, University Mustapha Stambouli of Mascara, Mascara, Algeria

Paramartha Dutta Visva-Bharati University, Santiniketan, India

Sima Ghosh Department of Civil Engineering, National Institute of Technology Agartala, Agartala, Tripura, India

Samir Ghouali Faculty of Sciences and Technology, Mustapha Stambouli University, Mascara, Algeria;
STIC Laboratory, Faculty of Engineering, University of Tlemcen, Tlemcen, Algeria

M. M. Hasan Department of Mathematics, Comilla University, Cumilla, Bangladesh;
Department of Applied Mathematics, University of Dhaka, Dhaka, Bangladesh

M. M. Hossain Department of Applied Mathematics, University of Dhaka, Dhaka, Bangladesh

M. S. Islam Department of Mathematics, Bangabandhu Sheikh Mujibur Rahman Science and Technology University, Gopalganj, Bangladesh

Sun Jamatia Department of Computer Science and Engineering, Tripura Institute of Technology, Narsingarh, Tripura, India

D. Loganathan Pondicherry Engineering College, Puducherry, India

Geetanjali Lohar Department of Civil Engineering, National Institute of Technology Agartala, Agartala, Tripura, India

Abhishek Majumder Department of Computer Science and Engineering, Tripura University, Agartala, Tripura, India

Swanirbhar Majumder Department of Information Technology, Tripura University, Agartala, Tripura, India

Sajib Mandal Department of Mathematics, Bangabandhu Sheikh Mujibur Rahman Science and Technology University, Gopalganj, Bangladesh

Boualem Merabet Faculty of Sciences and Technology, Mustapha Stambouli University, Mascara, Algeria

Subhasish Mohapatra Department of Computer Science & Engineering, Adamas University, Kolkata, India

Champa Nandi Department of Electrical Engineering, Tripura University, Agartala, Tripura, India

Swarup Nandi Department of Information Technology, Tripura University, Agartala, Tripura, India

Jayanta Pal Department of Information Technology, Tripura University, Agartala, Tripura, India

Komal Paul Department of Computer Science & Engineering, Adamas University, Kolkata, India

Anoop Kumar Prasad Computer Science and Engineering, Assam Science and Technology University, Royal School of Engineering and Technology, Guwahati, Assam, India

Abigail Rai Department of Computer Application, SMIT, Sikkim Manipal Institute of Technology, Sikkim, India

Srimanta Ray National Institute of Technology Agartala, Agartala, Tripura, India

Suvendra Kumar Ray Department of Molecular Biology and Biotechnology, Tezpur University, Tezpur, Assam, India

Alak Roy Department of Information Technology, Tripura University, Agartala, Tripura, India

Abhishek Roy Department of Computer Science & Engineering, Adamas University, Kolkata, India;
International Association of Engineers, Hong Kong, China;
Cryptology Research Society of India, ISI Kolkata, Kolkata, India

Bishwa Ranjan Roy Department of Computer Science, Assam University Silchar, Silchar, Assam, India

Apu Kumar Saha Department of Mathematics, National Institute of Technology Agartala, Agartala, Tripura, India

M. A. Samad Department of Applied Mathematics, University of Dhaka, Dhaka, Bangladesh

Nityananda Sarma Tezpur University, Tezpur, Assam, India

Santosh Kumar Satapathy Pondicherry Engineering College, Puducherry, India

Siddhartha Sankar Satapathy Department of Computer Science and Engineering, Tezpur University, Tezpur, Assam, India

Piyali Sen Department of Computer Science and Engineering, Tezpur University, Tezpur, Assam, India

Sushmita Sharma Department of Mathematics, National Institute of Technology Agartala, Agartala, Tripura, India

Smriti Kumar Sinha Department of Computer Science and Engineering, Tezpur University, Napaam, Tezpur, Assam, India

Abu Sufian University of Gour Banga, Malda, India

Debapriya Tula Department of Computer Science and Engineering, IIIT, Sri City, Chittoor, Andhra Pradesh, India

Design of an Industrial Internet of Things-Enabled Energy Management System of a Grid-Connected Solar–Wind Hybrid System-Based Battery Swapping Charging Station for Electric Vehicle

Somudeep Bhattacharjee and Champa Nandi

Abstract Increasing greenhouse gases imposes severe concern over the environment since it results in rising dangerous calamities of climate change in the form of flood, cyclone, the rise of sea level, and so on. By promoting renewable power generation and electric vehicles, greenhouse gas emissions can be reduced to a very low level. But both the solutions have some major disadvantages like the intermittency of renewable sources is very high and also electric vehicles need to be charged after traveling a fixed distance. This paper mainly provides a remedy for these disadvantages. In this study, a grid-connected solar–wind hybrid system-based battery swapping charging station for the electric vehicle is designed, which includes an IIoT (Industrial Internet of Things)-enabled energy management system to efficiently utilize and control the flow of energy of different sources. This study includes a twenty-four-hour case study analysis on Meghalaya, India, by utilizing the real-time data of solar radiation and wind speed of January month to check the feasibility and power generation capacity. The results of this analysis simply indicate that the IIoT-enabled energy management system is efficiently managing the energy from different renewable energy sources in the proposed hybrid system for supplying the load and for storing a fixed amount of energy in the battery for electric vehicle charging which shows that the overall hybrid system is feasible, profitable, and environmentally friendly.

Keywords Hybrid energy system · Climate change · Renewable energy · Industrial internet of things · Electric vehicle · Battery swapping charging station

---

S. Bhattacharjee · C. Nandi (✉)
Department of Electrical Engineering, Tripura University, Suryamaninagar, Agartala 799022, Tripura, India
e-mail: cnandi@tripurauniv.in

S. Bhattacharjee
e-mail: somudeeptit812@gmail.com

J. K. Mandal et al. (eds.), *Applications of Internet of Things*, Lecture Notes in Networks and Systems 137, https://doi.org/10.1007/978-981-15-6198-6_1

1 Introduction

One of the significant challenges of this world is air pollution, mainly due to increasing greenhouse gases in the environment. These greenhouse gases are responsible for rising dangerous calamities of climate change in the form of flood, cyclone, increasing global temperature, the rise of sea level, and so on [1]. Most of the countries of the world, including India, generate power depending upon coal and diesel feed thermal power plants [2]. The use of coal and diesel, to such a large extent, is the main reason for greenhouse gas emission. Thus, the utilization of renewable energy resources is a must and necessary for decreasing greenhouse gas emission, but renewable energy sources are highly intermittent [2, 3]. Another major cause of greenhouse gas emission is the extensive utilization of petrol and diesel vehicles [4–6]. In order to reduce the utilization of petrol and diesel vehicles, the utilization of electric vehicles needs to be increased [7–12]. An electric vehicle is mainly a substitute design of an automobile, which utilizes an electric motor to feed power to the car, with the electrical energy being supplied by a battery [4, 13–20]. Electric vehicles utilize electricity as fuel, so it does not emit any greenhouse gases, but it needs a charging station after traveling a fixed amount of distance for charging its battery [4]. This drawback of the electric vehicle can be easily solved by utilizing the concept of battery swapping. Battery swapping stations are those stations where an empty battery of an electric vehicle can be replaced by a fully charged battery of the same variety, which helps to minimize customer's tension about having sufficient power in the battery for the journey [13]. Petrol, diesel, or gas stations for ordinary vehicles can be easily converted into battery swapping stations as it requires a small investment in infrastructure, skilled workers who can replace the battery, and a reserve of fully charged batteries of the electric vehicle [14]. In addition, replacing an empty battery with a fully charged battery in an electric vehicle consumes less time as compared to the time required to charge the battery [14]. But batteries used in battery swapping stations must contain clean power generated from renewable energy sources; otherwise, it only creates one more reason for increasing greenhouse gas emissions in the environment. Therefore, both the solutions of decreasing greenhouse gas emissions had now arrived at one significant aim of solving the problem of the intermittent nature of renewable energy resources. The problem of intermittency of renewable energy sources can be easily solved by utilizing the concept of the hybrid energy system [2]. In the hybrid energy system, two or more energy sources are integrated to efficiently utilize and control the flow of energy from different sources [2].

Some papers are discussed here as a review related to the energy management system of the hybrid power plant. In [4], the authors propose an intelligent energy management controller of a grid-connected solar–wind-thermal-based hybrid energy system. The main aim of this hybrid system is to supply power to an electric vehicle charging station for reducing vehicle pollution. This research also includes a case study analysis using the real-time data of solar irradiance and wind speed of Delhi, India. One major limitation of the hybrid system in [4] is that the proposed energy management algorithm only considering the factor of power generation and not the factor of variation in load demand. In [2], a grid-connected solar–wind hybrid energy system is designed to deliver power in Egypt. In this work, the operation of the system is analyzed by making hourly energy balance calculations of a year. The energy flow to and from every element in the framework is determined by the contrast between the accessible and the demand energies. This research also includes an optimization case study analysis using the real-time data of solar irradiance and wind speed of the selected location. Results indicate that the load demand is fulfilled with the minimum levelized cost of energy and with no emission of greenhouse gases. One major limitation of the hybrid system in [2] is that the proposed energy management algorithm only focuses on reducing pollution from power plants and not on vehicle pollution. Due to the high renewable power generation possibility, the proposed hybrid system is capable enough to install a charging station for the electric vehicle in the selected location, which not only increases the yearly earnings of the hybrid system but also reduces vehicle pollution of that area. The two mentioned demerits of the above papers are trying to be solved in this research study.

In this study, a grid-connected solar–wind hybrid energy system-based battery swapping charging station for the electric vehicle is designed for fulfilling the load demand and for storing a fixed amount of energy in the battery for electric vehicle charging. The stored energy is aimed to utilize for electric vehicle charging using the battery swapping concept. The power output from renewable sources changes; therefore, they have to be regulated, so a backup grid connection is provided that can take excess power as well as supply required power during low power generation. In order to set up a proficient energy management approach, an IIoT (Industrial Internet of Things)-enabled energy management system is designed, which takes the decision based on the condition of the electric load, voltage level, battery state of charge, and energy generation. In this study, it is considered that the data acquisition system of battery and electric load consists of multiple IIoT devices (sensors) that use the Internet to share valuable information to the energy management system regarding electric load demand and battery state of charge [21, 22]. The information, therefore, received would be used by IIoT-enabled energy management system for deciding its control actions. The IIoT-enabled energy management system would utilize an implanted energy management algorithm to take decisions, which are then used to manage the utilization of the power outputs to maintain load-supply power balance. The accurate intention of this work is to formulate and apply an energy management algorithm for providing consistent power to the load and for storing a fixed amount of energy in the battery for electric vehicle charging in the hybrid energy system by designing an IIoT-enabled energy management system. This type of energy management system mainly shows its efficacy in managing the intermittency of renewable energy sources. This study also includes a twenty-four-hour case study analysis on Meghalaya, India, by utilizing the real-time data of solar radiation and wind speed of January month to check the feasibility and power generation capacity.

The rest of the article is organized as follows. Section 2 describes the modeling and simulation of the grid-connected solar–wind hybrid energy system-based battery swapping charging station. Section 3 demonstrates the simulation results of twenty-four-hour case study analysis on Meghalaya, India, by utilizing the real-time data of solar radiation and wind speed of January month. Section 4 provides the conclusion of the research analysis.

2 Modeling and Simulation of Grid-Connected Solar–Wind Hybrid Energy System-Based Battery Swapping Charging Station

A simulation model of the grid-connected solar–wind hybrid energy system-based battery swapping charging station for the electric vehicle, including the energy management system, is designed, which is shown in Fig. 1. This hybrid system consists of five main parts: PV array system, wind farm, grid, and electric load, electric vehicle charging station, and energy management system.

2.1 PV Array System

In the PV array system module, the specifications of the PV array system are as follows: Total area of PV array system is 5000 m^2, total rated capacity of the PV array system with MPPT is 1 MW, solar panel efficiency is 15%, and performance ratio is 0.75. The PV array system is used to convert sunlight into DC electric energy, which is increased using maximum power point tracker (MPPT) of efficiency 86% and send to the energy management system.

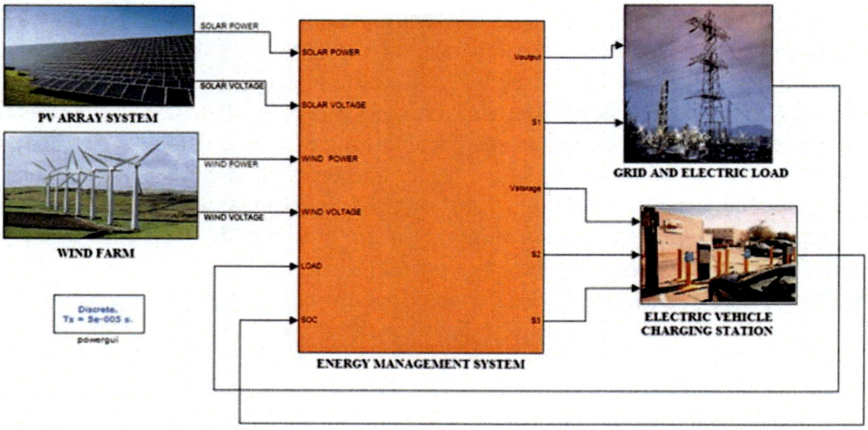

Fig. 1 Grid-connected solar–wind hybrid energy system-based battery swapping charging station for the electric vehicle including the energy management system

2.2 *Wind Farm*

In the wind farm module, the specifications of wind farm module are as follows: The rated power of one windmill is 10 kW, the power coefficient is 0.46, the efficiency of wind turbine generator is 90%, the wind turbine swept area is 38.4845 m^2, the cut-in speed is 2.7 m/s, the cut-out speed is 20 m/s, the number of windmills in the wind farm is 40, the density of air is 1.225 kg/m^3, the rated wind speed is 10.082 m/s, the hub height of wind turbine is 52 m, the anemometer height is 50 m, and the power-law exponent is (1/7). The wind farm system is used to convert winds into AC electric energy, which is converted into DC electric energy using the rectifier of efficiency 80%, and then it is increased using MPPT of efficiency 80% and send to the energy management system.

2.3 *Grid and Electric Load*

The grid and electric load module simply represent a DC electric grid where the input voltage from renewable sources must be above 440 V; otherwise, problems related to grid stability arise. In addition, to transmit excess energy out of a hybrid energy system back onto the grid, the voltage must be increased above the grid voltage. So V_{GRID} is the minimum input DC voltage (i.e., 440 V). The IIoT-enabled energy management system takes care of this issue by monitoring V_{GRID}. In this DC grid, DC input voltage is converted into AC voltage using IGBT two-level inverter, and then, this AC voltage is filtered using LC filter to remove harmonics. The filtered AC voltage is used to feed 50 Hz electric load of 100 kW. Since this load demand is continuously monitored by the energy management system so in case of load fluctuations, no problem would arise. For emergency conditions, a 50 Hz voltage source (representing AC grid) is connected near to the electric load for maintaining stability. This AC grid is mainly integrated either to consume excess energy or to feed the required energy to the grid during requirement.

2.4 *Electric Vehicle Charging Station*

In the electric vehicle charging station module, a battery bank of lithium-ion battery is present, which is used to store energy in the form of voltage. The power of the battery is dissipated through the load resistor. The rated capacity of the battery bank is 2400 Ah, and the nominal voltage is 300 V. The state of charge (SOC) of the battery is continuously monitored by the IIoT-enabled energy management system to prevent overcharging. The stored energy is aimed to utilize for electric vehicle charging using the battery swapping concept. Hence, it is necessary to keep the input voltage of the electric vehicle charging station module more than the nominal voltage of battery;

otherwise, discharging occurs in the reverse direction and so V_{NB} is the nominal voltage of the battery bank (i.e., 300 V) of the electric vehicle charging station. The IIoT-enabled energy management system takes care of this issue by monitoring V_{NB}. Based upon this concept, the stored energy is aimed to use to charge small rated capacity batteries of electric vehicles and then transported to all those places where it can sell to consumers.

2.5 *Energy Management System*

The energy management system is also known as IIoT-enabled energy management system. In this study, it is considered that the data acquisition system of battery and electric load consists of multiple IIoT devices (sensors) that use the Internet to share valuable information to the energy management system regarding electric load demand (LOAD) in watt and battery state of charge (SOC) in %. The information, therefore, received would be used by IIoT-enabled energy management system for deciding its control actions. The IIoT-enabled energy management system would utilize an implanted energy management algorithm to take decisions, which are then used to manage the utilization of the power outputs to maintain load-supply power balance. The flowchart of the energy management algorithm is shown in Fig. 2. Based upon the energy management algorithm, the IIoT-enabled energy management system decides working actions. For its operation, IIoT-enabled energy management system takes the input of solar DC power P_{SOLAR} (in watt), wind DC power P_{WIND} (in watt), solar DC voltage V_{SOLAR} (in volt), wind DC voltage V_{WIND} (in volt), LOAD (in watt), and battery state of charge (in %). Here, LOAD indicating the electric load demand, which the proposed hybrid system needs to be fulfilled. In addition, V_{GRID} (440 V) and V_{NB} (300 V) are already set in IIoT-enabled energy management system on the basis of the specifications of the grid and electric load module and electric vehicle charging station module. The working of IIoT-enabled energy management system based on energy management algorithm is divided into four different cases: In the first case, initially, the solar power P_{SOLAR} is checked that it is more than or equal to LOAD or not and if it is more than or equal to LOAD, then the solar voltage V_{SOLAR} is sent to the grid and electric load module. At that time, wind voltage V_{WIND} is sent to the electric vehicle charging station module for charging the battery bank. In this case, V_{OUTPUT} is equal to V_{SOLAR} and $V_{STORAGE}$ is equal to V_{WIND}. If V_{OUTPUT} is more than or equal to V_{GRID} (440 V), then V_{OUTPUT} is used to fulfilled load demand and excess energy sent to the grid in the grid and electric load module. But if V_{OUTPUT} is less than V_{GRID} (440 V), then the supply of V_{OUTPUT} is disconnected from the grid and electric load using a circuit breaker in the grid and electric load module. In this situation, if the grid power is available, then it is used to fulfill the load demand; otherwise, load shedding occurs. If $V_{STORAGE}$ is more than or equal to V_{NB} (300 V) and if SOC is less than or equal to 80%, then $V_{STORAGE}$ is used to charge the battery bank in the electric vehicle charging station module; otherwise, the supply of $V_{STORAGE}$ is disconnected from the battery bank and connected to dump load in

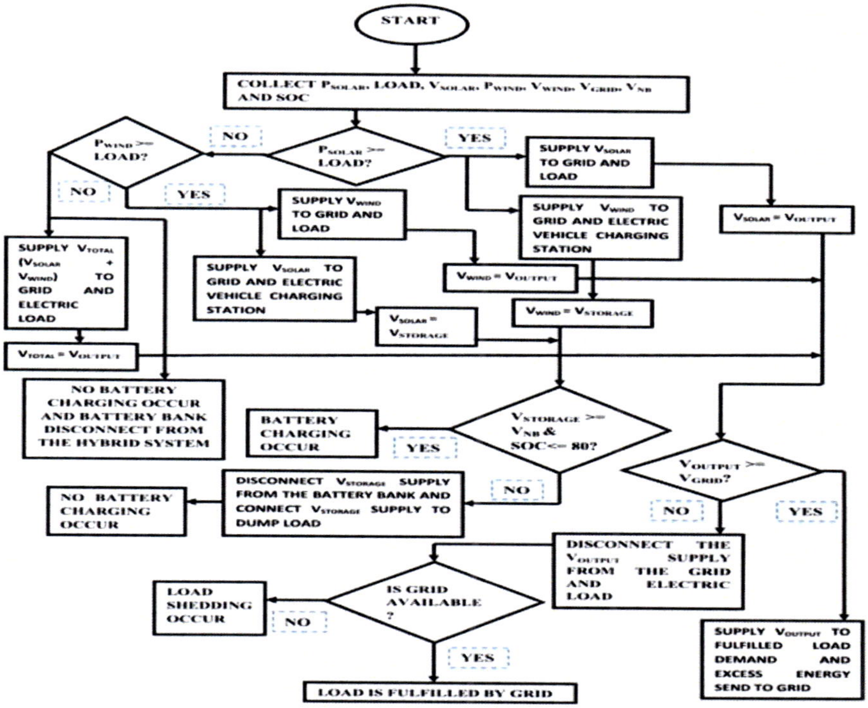

Fig. 2 Flowchart of the energy management algorithm

the electric vehicle charging station module. In this situation, no battery charging occurs. In the second case, if the solar power P_{SOLAR} less than LOAD, then the wind power P_{WIND} is checked that it is more than or equal to LOAD or not and if it is more than or equal to LOAD, then the wind voltage V_{WIND} is sent to the grid and electric load module. At that time, solar voltage V_{SOLAR} is sent to the electric vehicle charging station module for charging the battery bank. In this case, V_{OUTPUT} is equal to V_{WIND}, and $V_{STORAGE}$ is equal to V_{SOLAR}. If V_{OUTPUT} is more than or equal to V_{GRID} (440 V), then V_{OUTPUT} is used to fulfilled load demand and excess energy sent to the grid in the grid and electric load module. But if V_{OUTPUT} is less than V_{GRID} (440 V), then the supply of V_{OUTPUT} is disconnected from the grid and electric load using a circuit breaker in the grid and electric load module. In this situation, if the grid power is available, then it is used to fulfill the load demand; otherwise, load shedding occurs. If $V_{STORAGE}$ is more than or equal to V_{NB} (300 V) and if SOC is less than or equal to 80%, then $V_{STORAGE}$ is used to charge the battery bank in the electric vehicle charging station module; otherwise, the supply of $V_{STORAGE}$ is disconnected from the battery bank and connected to dump load in the electric vehicle charging station module. In this situation, no battery charging occurs.

In the third case, if both the solar power P_{SOLAR} and wind power P_{WIND} are individually less than LOAD, then the total power P_{TOTAL} ($P_{SOLAR} + P_{WIND}$) is calculated and check that it is more than or equal to LOAD or not and if it fulfills this condition, then the total voltage V_{TOTAL} ($V_{SOLAR} + V_{WIND}$) is sent to the grid and electric load module. At that time, no voltage sent to the electric vehicle charging station module for charging the battery bank. In this case, V_{OUTPUT} is equal to V_{TOTAL} and $V_{STORAGE}$ is equal to zero. If V_{OUTPUT} is more than or equal to V_{GRID} (440 V), then V_{OUTPUT} is used to fulfilled load demand and excess energy sent to the grid in the grid and electric load module. But if V_{OUTPUT} is less than V_{GRID} (440 V), then the supply of V_{OUTPUT} is disconnected from the grid and electric load using a circuit breaker in the grid and electric load module. In this situation, if the grid power is available, then it is used to fulfill the load demand; otherwise, load shedding occurs. Since $V_{STORAGE}$ is less than V_{NB} (300 V), the supply of $V_{STORAGE}$ is disconnected from the battery bank and connected to dump load in the electric vehicle charging station module. In this situation, no battery charging occurs. In the fourth case, if the total power P_{TOTAL} ($P_{SOLAR} + P_{WIND}$) is less than LOAD, then if the grid power is available, then it is used to fulfill the load demand otherwise load shedding occur. In this case, the hybrid system is disconnected from the battery bank, and no battery charging occurs.

3 Case Study in Meghalaya, India

3.1 Solar Radiation, Clearness Index, and Wind Speed

The latitude and longitude of the chosen location of Meghalaya, India, are 25°28.0′ N and 91°22.0′ E. Figure 3 indicates the monthly average information of daily solar radiation in kWh/m^2/day and clearness index. Figure 4 indicates the monthly average information of wind speed in m/s. Figures 3 and 4 are obtained from the real-time data of solar radiation, clearness index, and wind speed of the chosen location, which are taken from the database of NASA Prediction of Worldwide Energy Resources [23]. The chief reason for picking up this location is its maximum possibility of renewable energy production, which is shown in Figs. 3 and 4.

In most of the states of Northeast India, including Meghalaya, the transportation sector is the main contributor to greenhouse gas emissions due to consuming diesel and petrol. The total emissions of 2012–13 baseline year of Meghalaya were 2.96 million tons CO_2 equivalent, in which the power sector contributed 56,238 tons CO_2 equivalent (3.53%), and the transport sector contributed 1,004,106 tons CO_2 equivalent (62.96%) [24, 25]. Therefore, these emissions can be reduced by utilizing the possibility of renewable energy production.

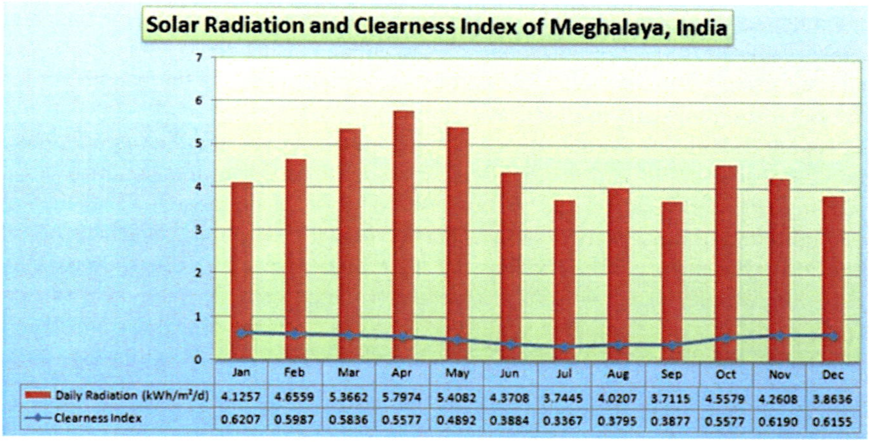

Fig. 3 Monthly average information of daily solar radiation and clearness index

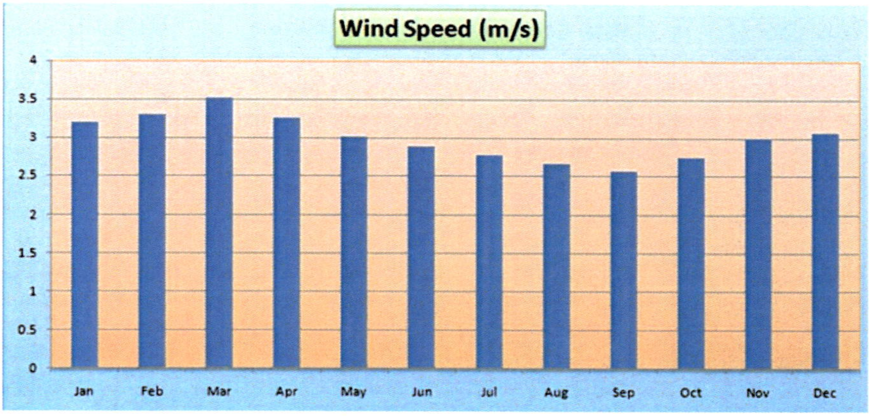

Fig. 4 Monthly average information of wind speed

3.2 Renewable Energy Generation Analysis

In order to analyze the feasibility and power generation capacity of the proposed hybrid system by means of the IIoT-enabled energy management system at various conditions, a 24-hour duration case study analysis on Meghalaya, India, is done by means of the real-time data of solar radiation and wind speed of one day of January month of the chosen location. The analysis based on the results obtained from the simulation model of the proposed hybrid system by using real-time data. The hourly results of solar power, solar voltage, wind power, wind voltage, output voltage, and storage voltage obtained from the simulation model of the proposed hybrid system with the real-time data of solar radiation and wind speed of the chosen location are shown in Table 1. The output voltage is the voltage of the energy sent to the grid, and the storage voltage is the voltage of the energy sending to the battery bank of the electric vehicle charging station.

Figure 5 indicates the hourly power utilization scenario in which hourly excess power sending to the grid, required power taking from the grid, and storage power sending to the electric vehicle charging station in watt are present. With the help of Fig. 5 and Table 1, it is easy to understand all the situations that the proposed hybrid system mainly faces and how the situation is managed. From 0:00 to 4:00 h time duration (12:00 a.m.–4:00 a.m.), both the solar power and wind power are zero. During this time, load demand is fulfilled by the grid, and the charging of the battery bank of the electric vehicle charging station not occurs. From 4:00 to 6:00 h time

Table 1 Hourly results of the hybrid energy system

Time	Solar radiation (kW/m^2)	Wind speed (m/s)	Solar power (W)	Solar voltage (V)	Wind power (W)	Wind voltage (V)	Output voltage (V)	Storage voltage (V)
0:00	0	2.270178	0	0	0	0	0	0
1:00	0	2.030897	0	0	0	0	0	0
2:00	0	2.096759	0	0	0	0	0	0
3:00	0	2.448828	0	0	0	0	0	0
4:00	0	2.874705	0	0	13,580	1165	1165	0
5:00	0	3.151139	0	0	17,890	1337	1337	0
6:00	0.050017	4.032377	52,330	2288	37,480	1936	4224	0
7:00	0.103069	4.931006	107,800	3284	68,540	2618	3284	2618
8:00	0.155931	5.395521	163,100	4039	89,790	2996	4039	2996
9:00	0.256246	5.193318	268,100	5178	80,070	2830	5178	2830
10:00	0.048092	5.903589	50,320	2243	80,070	2830	5073	0
11:00	0.031911	4.890349	33,390	1827	66,860	2586	4413	0
12:00	0.196842	6.265872	205,900	4538	140,600	3750	4538	3750
13:00	0.156045	7.846434	163,300	4041	276,100	5255	4041	5255
14:00	0.054842	7.417167	57,380	2395	233,300	4830	7225	0
15:00	0.022347	8.431877	23,380	1529	342,700	5854	7383	0
16:00	0.013633	7.466731	14,260	1194	238,000	4878	6072	0
17:00	0	6.14351	0	0	132,500	3641	3641	0
18:00	0	5.902899	0	0	117,600	3429	3429	0
19:00	0	5.327314	0	0	86,430	2940	2940	0
20:00	0	3.789236	0	0	31,100	1764	1764	0
21:00	0	4.813469	0	0	63,750	2525	2525	0
22:00	0	4.820792	0	0	64,040	2531	2531	0
23:00	0	4.682683	0	0	58,690	2423	2423	0

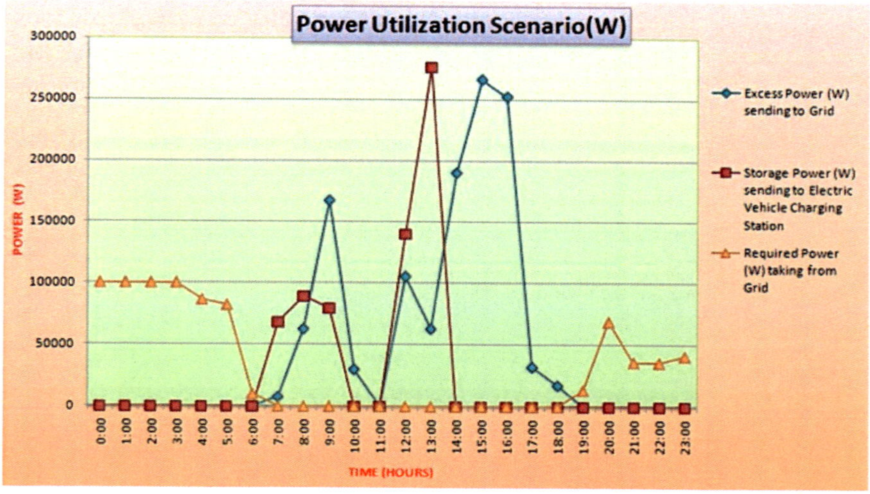

Fig. 5 Hourly power utilization scenario in watt

duration (4:00 a.m.–6:00 a.m.), the wind power is only available which is lower than the load demand, so load demand is fulfilled by the combined power of wind and grid, and in this situation, the charging of battery bank of electric vehicle charging station not occurred. From 6:00 to 8:00 h time duration (6:00 a.m.–8:00 a.m.), both the solar and wind power generation are rising, but the total power is less than the load demand, so load demand is fulfilled by the combined power of solar, wind, and grid. In this situation, the charging of the battery bank of the electric vehicle charging station not occurred.

From 8:00 to 17:00 h time duration (8:00 a.m.–5:00 p.m.), solar power generation raised above the load demand, so it is sent to fulfill the load demand and excess solar energy sent to the grid. At that time, wind power was sent for storage in the battery bank of the electric vehicle charging station. In this time period, solar power generation sometimes reduces below load demand, so at that time, the total power of solar and wind sources was used to fulfill the load demand, and in this situation, the charging of battery bank of electric vehicle charging station did not occur. From 17:00 to 0:00 h time duration (5: 00 p.m.–12:00 a.m.), the wind power is only available which is lower than the load demand, so load demand is fulfilled by the combined power of wind and grid, and in this situation, the charging of battery bank of electric vehicle charging station not occurs. The overall production and consumption summary of the electrical energy of the proposed hybrid system of one day of January month of the chosen location is shown in Table 2. This table indicates that the total generation of renewable energy is high as well as the total surplus energy sending to the grid is greater than the total required energy taking from the grid, which shows that the overall hybrid system provides a profitable business.

Table 2 Electrical production and consumption summary of the hybrid system on Meghalaya, India

Production summary of the hybrid system		
Component	Production (Wh/day)	Percent (%)
PV array system	1,139,260	27.43182135581956
Wind farm	2,239,090	53.91422228429158
Total required energy taking from grid	774,710	18.65395635988885
Total	4,153,060	100

Consumption summary of the hybrid system		
Component	Consumption (Wh/day)	Percent (%)
AC primary load	2,400,000	56.42995866505528
Total surplus energy sending to grid	1,197,960	28.16701386766234
Total storage energy sending to electric vehicle charging station	655,100	15.40302746728238
Total	4,253,060	100

4 Conclusion

The overall conclusion of the analysis is that the IIoT-enabled energy management system efficiently manages the energy from different renewable energy sources in the proposed hybrid system for supplying the load and for storing a fixed amount of energy in the battery for electric vehicle charging. It successfully stores a good amount of energy in the battery bank of the electric vehicle charging station, which can be used for electric vehicle charging using the battery swapping concept. Since the utilization of electric vehicle mainly occurs during day time, the electric vehicle charging station is successfully able to charge a maximum number of electric vehicles with this stored energy. In addition, it successfully manages the fluctuation of renewable energy with no emission of greenhouse gases. The results of the proposed hybrid system indicate that the system is feasible, profitable, and environmentally friendly. This work not only increases the utilization of renewable energy in fulfilling load demand but also promotes electric vehicles in the chosen location if implemented. This work can be further extended by integrating a biogas power plant in the proposed hybrid system.

References

1. Climate Risk Assessment and Management: Tamil Nadu State Planning Commission and Regional Integrated Multi Hazard Early Warning Systems. https://www.unescap.org/sites/default/files/Climate%20risk%20assessment%20tools%20for%20development%20planning%20by%20Sugato%20Dutt.pdf. Last accessed 1 Mar 2019
2. Nandi, C., Bhattacharjee, S., Chakraborty, S.: Climate change and energy dynamics with solutions: a case study in Egypt. In: Qudrat-Ullah, H., Kayal, A. (eds.) Climate Change and Energy

Dynamics in the Middle East. Understanding Complex Systems, pp. 225–257. Springer, Berlin (2019)
3. Islam, M.A., Hasanuzzaman, M., Rahim, N.A., Nahar, A., Hosenuzzaman, M.: Global Renewable Energy-Based Electricity Generation and Smart Grid System for Energy Security, pp. 1–13. The Scientific World Journal, Hindawi Publishing Corporation, London (2014)
4. Bhattacharjee, S., Nandi, C., Reang, S.: Intelligent energy management controller for hybrid system. In: 3rd IEEE International Conference for Convergence in Technology (I2CT), pp. 1–7. IEEE, Pune, India (2018)
5. Al Wahedi, A., Bicer, Y.: Assessment of a Stand-alone Hybrid Solar and Wind Energy-Based Electric Vehicle Charging Station with Battery, Hydrogen and Ammonia Energy Storages, Energy Storage, pp. 1–17 (2019)
6. Huang, P., Ma, Z., Xiao, L., Sun, Y.: Geographic Information System-assisted optimal design of renewable powered electric vehicle charging stations in high-density cities. Appl. Energy **255**, 1–12 (2019)
7. Domínguez-Navarro, J.A., Dufo-López, R., Yusta-Loyo, J.M., Artal-Sevil, J.S., Bernal-Agustín, J.L.: Design of an electric vehicle fast-charging station with integration of renewable energy and storage systems. Int. J. Electr. Power Energy Syst. **105**, 46–58 (2019)
8. Dorotić, H., Doračić, B., Dobravec, V., Pukšec, T., Krajačić, G., Duić, N.: Integration of transport and energy sectors in island communities with 100% intermittent renewable energy sources. Renew. Sustain. Energy Rev. **99**, 109–124 (2019)
9. Badea, G., Felseghi, R.A., Varlam, M., Filote, C., Culcer, M., Iliescu, M., Răboacă, M.: Design and simulation of romanian solar energy charging station for electric vehicles. Energies **12**(1, 74), 1–16 (2019)
10. Lee, Y., Hur, J.: A simultaneous approach implementing wind-powered electric vehicle charging stations for charging demand dispersion. Renew. Energy **144**, 172–179 (2019)
11. Esfandyari, A., Norton, B., Conlon, M., McCormack, S.J.: Performance of a campus photovoltaic electric vehicle charging station in a temperate climate. Sol. Energy **177**, 762–771 (2019)
12. Kumar, V., Teja, V.R., Singh, M., Mishra, S.: PV Based Off-Grid Charging Station for Electric Vehicle, IFAC Workshop on Control of Smart Grid and Renewable Energy Systems (CSGRES 2019), vol. 52, no. 4, pp. 276–81 (2019)
13. Sarker, M.R., Pandzic, H., Ortega-Vazquez, M.A.: Electric vehicle battery swapping station: business case and optimization model. In: International Conference on Connected Vehicles and Expo (ICCVE), pp. 289–294. IEEE, Las Vegas, NV, USA (2013)
14. Bhattacharjee, S., Batool, S., Nandi, C., Pakdeetrakulwong, U.: Investigating electric vehicle (EV) charging station locations for Agartala, India. In: 2nd International Conference of Multidisciplinary Approaches on UN Sustainable Development Goals (UNSDGs), pp. 28–29. Bangkok, Thailand (2017)
15. Liu, L., Kong, F., Liu, X., Peng, Y., Wang, Q.: A review on electric vehicles interacting with renewable energy in smart grid. Renew. Sustain. Energy Rev. **51**, 648–661 (2015)
16. Richardson, D.B.: Electric vehicles and the electric grid: a review of modeling approaches, Impacts, and renewable energy integration. Renew. Sustain. Energy Rev. **19**, 247–254 (2013)
17. Dallinger, D., Wietschel, M.: Grid integration of intermittent renewable energy sources using price-responsive plug-in electric vehicles. Renew. Sustain. Energy Rev. **16**(5), 3370–3382 (2012)
18. Ma, Z., Callaway, D.S., Hiskens, I.A.: Decentralized charging control of large populations of plug-in electric vehicles. IEEE Trans. Control Syst. Technol. **21**(1), 67–78 (2013)
19. Tuttle, D.P., Baldick, R.: The evolution of plug-in electric vehicle-grid interactions. IEEE Trans. Smart Grid **3**(1), 500–505 (2012)
20. Yong, J.Y., Ramachandaramurthy, V.K., Tan, K.M., Mithulananthan, N.: A review on the state-of-the-art technologies of electric vehicle, its impacts and prospects. Renew. Sustain. Energy Rev. **49**, 365–385 (2015)
21. Bhattacharjee, S., Nandi, C.: Implementation of industrial internet of things in the renewable energy sector. In: Mahmood, Z. (ed.) The Internet of Things in the Industrial Sector, Computer Communications and Networks, pp. 223–259. Springer, Berlin (2019)

22. Madakam, S., Ramaswamy, R., Tripathi, S.: Internet of Things (IoT): a literature review. J. Comput. Commun. **3**(5), 164–173 (2015)
23. NASA Prediction of Worldwide Energy Resources. https://power.larc.nasa.gov/. Last accessed 15 June 2018
24. Jamir, T., De, U.S.: Trend in GHG emissions from northeast and west coast regions of India. Environ. Res. Eng. Manage. **1**(63), 37–47 (2013)
25. Carbon Footprint Study- Meghalaya State. https://meghalayaccc.org/wp-content/uploads/2019/03/Carbon-Footprint-Meghalaya-Report.pdf. Accessed 27 Aug 2019

Modeling and Implementation of Advanced Electronic Circuit Breaker Technique for Protection

Tushar Kanti Das, Rajesh Debnath, and Sangita Das Biswas

Abstract The following paper narrates a microcontroller-based system which is an advanced electronic circuit breaker that designed for voltage fluctuation, frequency fluctuation, short circuit, overload, and residual leakage current. The advanced circuit breaker announces various watchful parameters that users get information other then any smart energy device during any electrical fault-based accident. During twenty-first century, many IoT-based energy monitoring and control projects are done. This project has also on features of smart energy monitoring system in coordination with web server-based IoT model. However, this project can be initiated for the protection scheme of household service as well as protective model of smart power system [1, 2]. Nowadays, power system is dealing with high-voltage alternating current (HVAC) and extra high-voltage current (EHVC). For making high-voltage circuit breaker and protective devices, special attention should be taken for designing such equipment. The circuit breaker technique is used in this paper and can be installed in the protection scheme to make a fault-free power system and also IoT-enabled smart power system. A hardware prototype model is designed using Arduino microcontroller to make this project a successful one.

Keywords Advanced circuit breaker · Residual current leakage · Energy monitoring · Arduino · Internet of things

T. K. Das
Electrical Engineering, Techno College of Engineering Agartala, Madhuban, Tripura, India
e-mail: tushar.kd@yahoo.com

R. Debnath (✉) · S. Das Biswas
Department of Electrical Engineering, Tripura University, Suryamaninagar, Agartala 799022, Tripura, India
e-mail: rajdb.16@gmail.com

S. Das Biswas
e-mail: sdasbiswas@tripurauniv.in

J. K. Mandal et al. (eds.), *Applications of Internet of Things*, Lecture Notes in Networks and Systems 137, https://doi.org/10.1007/978-981-15-6198-6_2

1 Introduction

In earlier times, the circuit breaker evaluation is survey and it should be given special important that in preliminary phase of electrical application. Nowadays, uses of electrical energy are increased day by day. To prevent various faults, various types of CB are used. Everyone cannot use freely all types of circuit breaker for protect their certain system. Except RCCB the circuit breaker used here does not get any way to realize the system operational or faulty [3]. Conventional circuit breakers are designed by alloy-based thermal rocker arm tripping circuit. The electronic circuit breaker mainly making by the combination of automated switch and controller, controlled using the evaluation found from the load. It is designed in this way that its close off the power supply section when there any abnormal condition occurs which will be not acceptable for our domiciliary [4]. The circuit breaker differs to the conventional ones and also attach with proper actuator having an automatic switch. At basic condition, the circuit is normally closed (NC). After giving some signal from controller section, the NC circuits are automatically closed that means NC become normally open [NO] by the actuator. After clearing the section from fault if controller sends the green signal to actuator the circuit gets normally closed [5]. For the following project, at first the literature survey has been done on how the device needs to be modeled; then, a virtual circuit model has been made and simulated by Proteus 8 simulating software. After successful testing of the circuit, a hardware model has been implemented and the result has brought out using three different loads in the laboratory along with the help of a freeware IoT platform web server, i.e., (ThinkSpeak.com).

2 Literature Review

Energy management includes arrangement and performance of energy production and energy-consuming unit [6]. In this paper, we have done detailed study in consuming side energy monitoring, which is also one kind of management. One of the best things is that without measuring any data, it is quite impossible to manage anything [7, 8]. The fact is that for process scheduling, metering, and billing purpose, the monitoring of the energy is done. Main aim is to be in the field of management that discusses various drawbacks of the existing system and also using new generation technology making an advanced system. Advanced includes the real-time error less system. Introduction of the management technology in the field of industry has been used for recognizing the energy consumed by various equipments [9]. The power quality and energy management of the whole system is can be done by adapting intelligent energy meter [10].

Monitoring of the energy for valuable maters gives a clear scenario about the status of usage of the energy [11]. In this vision, the huge advancement of the information technology aims for the smart life achievement. In this paper, Internet of things technology is used as communication channel. The Internet of things is the recent trending topic for improving our quality life [12]. In the field of education, agriculture, medical, health care and every field IoT work are done tremendously, and it is also going to be improved day by day. So the contribution level of IoT in the field of conventional energy source is gaining in massive amount [13, 14].

For protecting the electrical system and equipments from various faults circuit breakers are used. From oil circuit breaker to electronic circuit breaker, lot of things changes, but the basic function remains same to protect the system [15]. Circuit breakers are the punctuate system which opens the circuit or close it as per their system command. Miniature circuit breaker (MCB) is design to protect from the electrical fault especially short circuit and overloads for electrical equipment and also human, by using automatic electrical operated switch [16]. The current set value is 100 A with over current protection. But molded case circuit breaker (MCCB) is a protective-type device used for a high-rating current and wide-range voltage [3]. As an earth leakage protector, an earth leakage circuit breaker (ELCB) is used. It is a safety device using in electrical field with very high earth impedance for preventing shock. Residual current circuit breaker (RCCB) is a protective device that protects the system from detecting that current is not balanced in various phases and neutrals. That results in imbalance condition. The system identifies whether current flows through neutral or not and also any earth leakage fault is there or not. The following system also detects overload condition, voltage & frequency fluctuation [17]. This system serves an important aspect of energy management and energy conservation also smart circuit protection solution.

The objectives of this project are as follows:

- Installation of this device gives a complete monitoring of the electrical energy from any part of the world via Internet.
- Display real-time voltage, current, power, frequency, energy units consume, earth voltage, residual current, and also circuit breaker status.
- The device is also capable of detecting whether any earth leakage fault is there or not and any current is flowing through neutral or not, if yes then within a fraction of second the circuit should be tripped by a relay.
- This project also provides protection scheme from voltage fluctuation, frequency fluctuation, short circuit, overload, and residual leakage current.
- Thereby ensuring a safe use of electricity to the consumer and giving a smart protection scheme.
- Monitoring and control the circuit breaker status in offline as well as online from anywhere using Internet of things.

Voltage transformer is used to sense the voltage level, current transformer is used to load current measurement, and residual leakage current measurement one transformer is used which will be called zero current transformer. The main three transformers are interlinking with the main controller for various types of transformers are used voltage transformers, current transformer, and zero current transformer brunch. In this system, one kind fault also reduced by measuring frequency through count

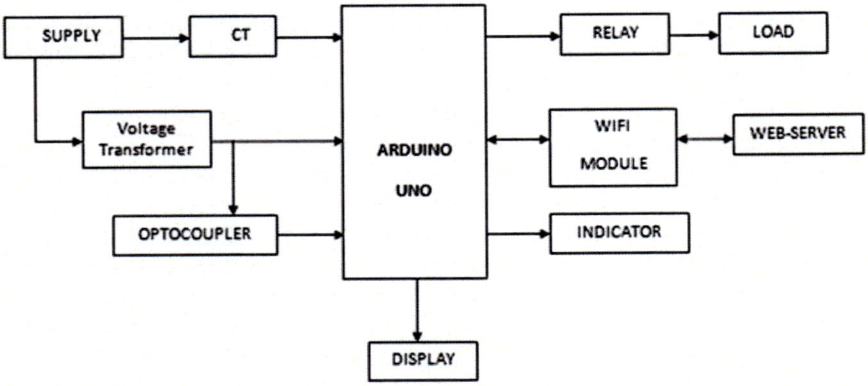

Fig. 1 Block diagram model of the proposed system

pulses of mains AC sine wave. Here this system also monitors the various system data, that's why one LCD is used. And here the last not a least one relay will be used for controlling the circuit. Controlling means for making and breaking the circuit. For this system, one DPDT, double-pole, double-throw relay is used (Fig. 1).

3 Proposed Model

The main perception of the project model is to provide energy monitoring and control with a secured protection scheme. In this proposed model, the microcontroller is the main unit. This will be used for various functions like as monitor and control. Arduino is an open-source networking platform for this project easy to simulate virtually as well as physically, so flexible. Current transformer is used for sensing the current for Arduino [18]. Connecting CT sensor with Arduino the output of the CT sensor is required condition. As a voltage sensor, voltage transformer used here is extremely accurately ratio step down transformer. Optocoupler is an electronic device that transfers electrical signal between two circuits which are electrically isolated. As a frequency sensor, this optocoupler is used. The Wi-Fi module here used is ESP8266. It is a Wi-Fi-enabled microchip consisting of a microcontroller and TCP/IP. A web server is a computer that serves or delivers data and services to end-users those which are connected over the Internet. Here for the project work, the web server that has been used is a private web server whose name is "Think Speak," and its database has been used to store content of the project. Relays are used for controlling the whole house electrical supply used in physically. If one unit, the whole system is sound; any fault, the microcontroller sends signal to relay for break the circuit. The display unit here used is type of LCD. LCD displays the real-time voltage, current, power, frequency, energy units consume, earth leakage voltage, residual current, and also the circuit breaker status. Indicator is used in the output side of the Arduino for giving alertness.

3.1 Flowchart of IoT-Enabled Proposed System

At first the controller will monitor the current, voltage, and frequency, and then, the system will check if there is any Wi-Fi network available or not. If available, the data will be sent to web server of "Think Speak" over Internet through Wi-Fi network. At a time, there is one program also running side by side that if the output of the sensors like voltage, current, frequency, or any value which more than predefined value set in the controller, the system will initiate for circuit breaking through its relay. And the controller is designed in such a way that if the circuit breaking parameters (such as earth fault, over voltage, frequency fluctuation, current through neutral) (Fig. 2).

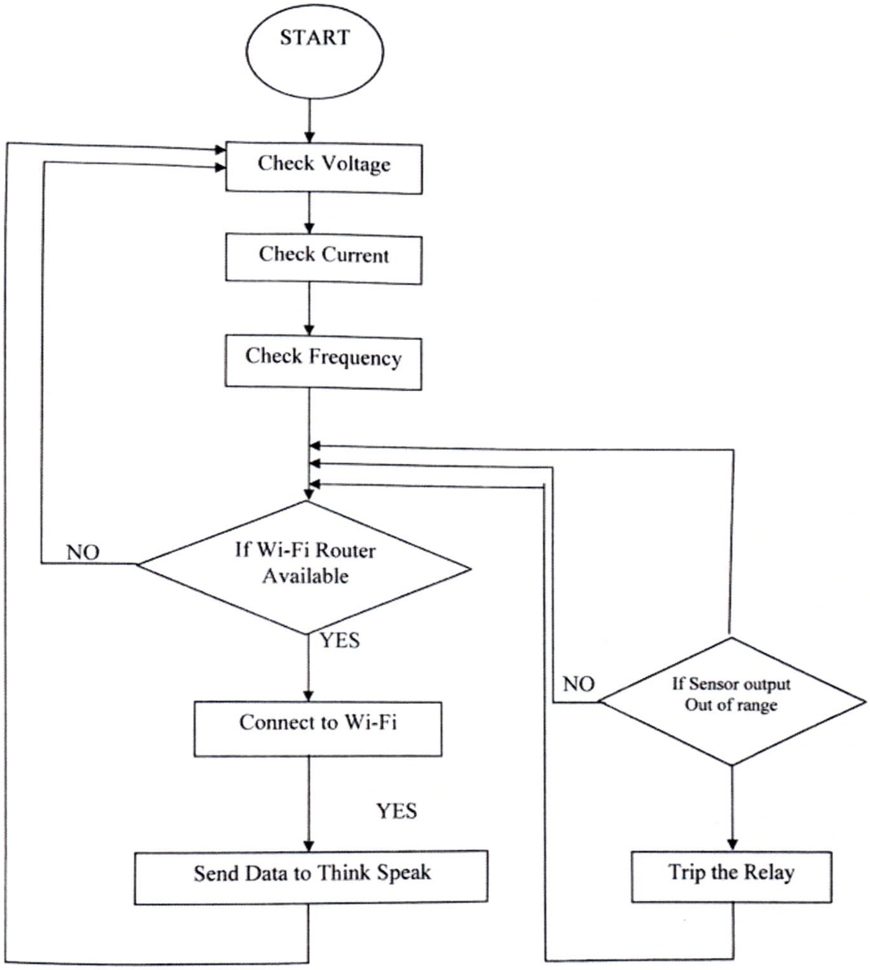

Fig. 2 Flowchart of IoT enabled the proposed system

4 Circuit Simulation

Simulink in Proteus software by using in Arduino Uno controller for monitoring the electrical load is done. Here Fig. 3 is showing the simulation part of monitoring portion of the Model; which shows the value of voltage, current, frequency and power. The Arduino pin no A1 is connecting with the CT or current sensor with the given AC source and the Arduino pin no A2 are connected with the VT with the supplying given AC voltage source. Here also the divider circuit also used for reducing the voltage level. Also capacitors are used in both parts because of the reducing harmonics and filtering.

The value of these parts we get from set the logical program by using electrical monitoring library in Arduino. In Arduino port no 3, we used the Opt-coupler circuit for measure the frequency. We know the frequency is known as the reciprocals the time period of the full-wave AC voltage. We first measure the time period. Opt-coupler used for detecting or counting the time period of how much time full-wave AC voltage is supplied. After that compiling the logical program in Arduino software, the simulation is done in Proteus 8 simulation software. All values we get from the CT and PT, and we seen it the 16 * 2 LCD unit by setting the logical program in Arduino that compiling the logical program in Arduino software, the simulation is

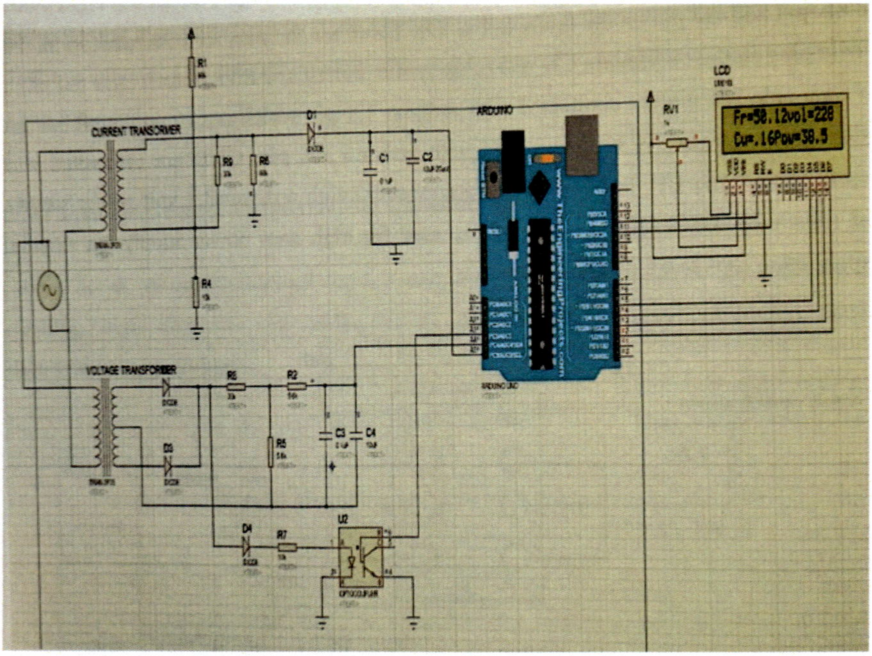

Fig. 3 Software simulation for monitoring

done in Proteus 8 simulation software [19]. All value we get from various sensor and parameter.

5 Hardware Model

There is nonstop progress in the field of electrical and electronic science. In the present project work, a system has been designed which monitors the electrical parameters like voltage, current, power, frequency also give the protection from the various faults and problems. This system designed in such a manner that help to monitor the system parameter as well as using this same system with the help of relay we can secure our places and equipment too. The one of major part of this project is IoT-based system. Monitor data which will be we get from the various sensor will be monitor from anywhere with the help of ESP8266 module and also the "Think Speak". "Think Speak" provides a free web server for project purpose we used this. We can also know the circuit breaker status through online (Fig. 4).

6 Result

This system is the advanced system of this new generation. One controller can make a several works done in a real time. We can say monitor and control using advanced electronic circuit breaker is a one kind of new generation path for automation industry. This model is implemented and successfully installed. Here, the results are shown in a LCD (Fig. 5) of the hardware model, where one 28 W load is connected across the system. Getting result from model display is further checked by conventional equipment (voltmeter, clamp meter), this is the only one condition (fault-free condition) where hardware behaves like that, at that times model monitors the frequency, supply voltage, load current, load power. At another condition, if frequency varies more within a short of time, this frequency affects the system voltage and load current also affects the equipment, at that time model display will indicated us in a message command from "Fr distortion" as a result using logical command of Arduino, the relay will be actuated to open the circuit. Again another condition if the voltage fluctuates or varies system may be affected at that's time the result in a command message form on display is "under vol" "over vol" and. Again one more condition if the load is high that means load current is high as compare to the normal user get command message as well as protection scheme to eliminated or cut the load from system through relay. In this way, our system becomes work as overload circuit breaker. So model gives the sufficient protection at different conditions using microcontroller.

Fig. 4 Hardware model of the proposed system

6.1 Earth Fault Detects System

Earth occurs connect the part which is not normally carrying any voltage to the earth return part. To reduce the risk of shock in the equipment metal earth, fault detector is necessary. This system is successfully worked in this manner. Sometime if the system is fault for line unbalanced load, and some voltages are flows through the earth metal wire. In this condition, earth leakage voltage was measured by the system, in a safe limit single phase 6.6 V is exceed it was worked and microcontroller give the signal to relay for disconnected the supply. ELCB has two types: One is voltage ELCB and another one is current ELCB. Our system worked as a voltage ELCB in a smart way. First monitor in a real-time earth leakage voltage and check it continuously, if exceed the limit it will be shown in the LCD at that time system will trip using

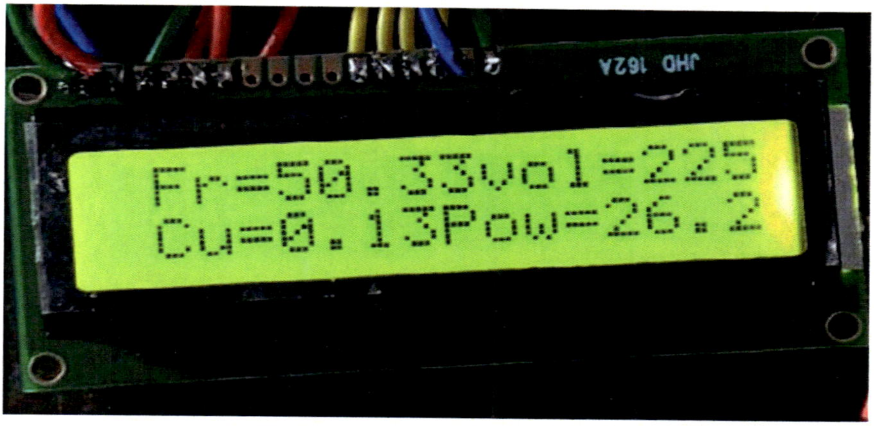

Fig. 5 Displaying the system electrical parameter

microcontroller and relay by the logical program. Given the result in a command message form in LCD is "earth fault detect."

6.2 *Residual Current Detects System*

Residual current is a current which will flow through the earth way and its similar as current earth leakage, sometimes if input current of system is not equal to outgoing current this model is activated. This model worked as good manner at a time of observation. This system is installed to prevent human from earth fault and protection for equipment. If the earth leakage current is occurred, the system is tripped the total circuit with leaving some message in model display.

6.3 *Experimental Result from Hardware Monitoring*

Table 1 shows the experimental result for monitoring load current and draws power. Here, the table has been formed using the reading taken from the working model connected with household 220 V, 50 Hz, load power factor 0.85 single-phase supply. The load current and the connected power have been taken into account as the actual voltage, and power is taken by conventional clamp meter and multimeter as well as equipment datasheet, and measured value is taken from nonconventional way that mean, taken it from model systems display. This value was pretty accurate for CFL bulb and LED bulb as well as incandescent bulb. This is the experimental result that means if load is resistive, inductive, or capacitive, this system worked properly.

Table 1 Experimental result for monitoring load current and draws power

Circuit parameters	For an electric bulb (pure resistive load)		For a CFL bulb		For a LED bulb	
	Actual value	Measured value	Actual value	Measured value	Actual value	Measured value
Current (Ampere)	0.181 A	0.170 A	0.069 A	0.065 A	0.050A	0.044 A
Load power (W)	40	38.9	13	12.7	9	8.5

6.4 IoT-Enabled Energy Monitoring Result

System is attached with web server through ESP8266 Wi-Fi module which continuously uploads the system parameter data to its server and enables the user to monitor the data over the Internet throughout worldwide. The device is connected with local Wi-Fi network with ESP8266. The web server here used is "Think Speak.com". It is a web server as well as an online Web site which enables the user to monitor and control its uploaded load data over Internet from anywhere. The "Think Speak.com" is freeware IoT hosting Web site, where user can create their own database for their need. Since the project is in its prototype stage, the "Think Speak.com" is well suited for this project due to free of cost. If the project is implemented commercially, a dedicated web server can be made for its secured operation. Here are some screenshots taken from Think Speak.com for the project's parameter data. The Field 1 chart taken from "Think Speak" server is the value of frequency with respect to time here shown it that (Fig. 6) on time 9:43:30, the value of frequency is 50.575 and after 1 min it was 50.55. Field 2 chart (Fig. 7) shows the data of power with respect to time. That means system parameter data is monitored through globe wide. In this way, we can say this model system is smart as well as interconnected devices over network connectivity.

7 Conclusions

Nowadays, power system safety and smart device or equipment plays vital role in commercial arena. Lots of work is going on the field of electrical and electronic engineering which based on IoT enabled. This system designed as electronics microcontroller-based smart circuit breaker which works as an all in one breaker. The developed model monitors the electrical parameters like voltage, current, frequency, power, and the fault status. The device is attached with a smart technique which owns the characteristics of earth faults or earth leakage faults, residual current circuit breaker, over load, voltage, and frequency fluctuation and giving the proper protection. This project found out to be a vital tool for energy efficient and energy management or building power management.

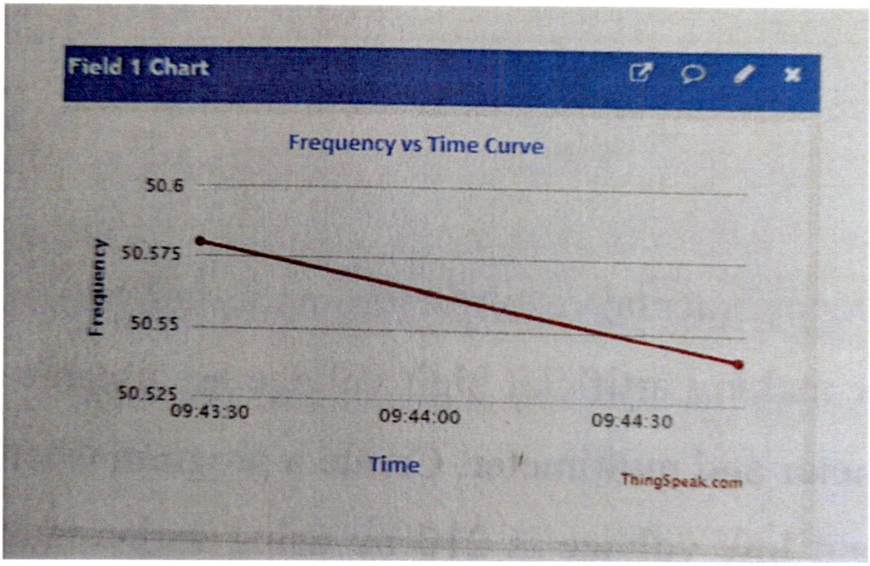

Fig. 6 Field 1 chart for frequency versus time

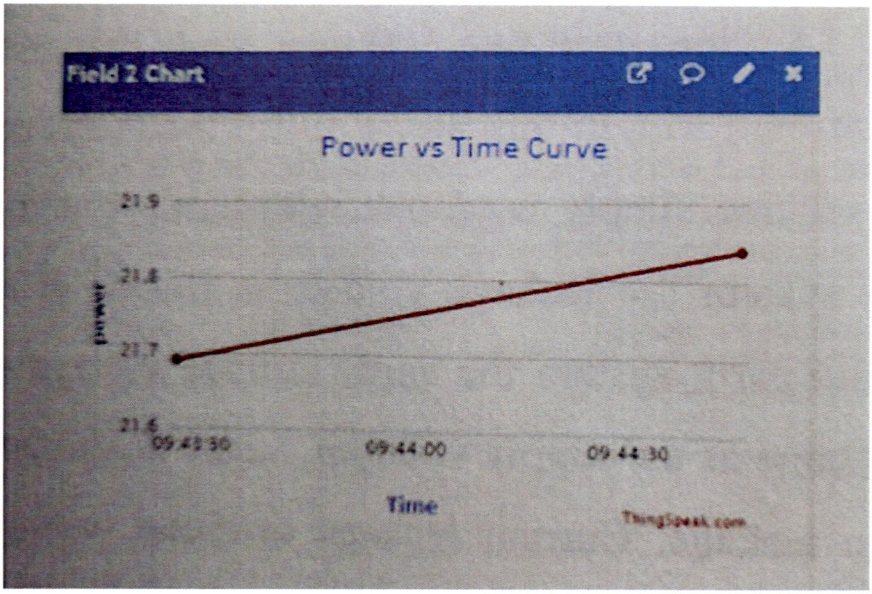

Fig. 7 Field 1 chart for power versus time

References

1. Li, W., Tan, X., Tsang, H.K.: Smart home energy management systems based on non-intrusive load monitoring. In: IEEE International Conference on Smart Grid Communications, Data Management, Grid Analytics, and Dynamic Pricing (2015)
2. Kodali, R.K., Jain, V., Bose, S., Boppana, L.: IOT Based Smart Security and Home Automation System, vol. 12 (2017). ISSN 0973-454442
3. Chen, D., Zhao Q., Chen F.: Adaptive residual current circuit breaker based on microcontroller. In: 2011 Second International Conference on digital Manufacturing and Automation, Human, China, 5–7 Aug 2011
4. Pallam, S.W., Usman, R., David, M., Luka, M.K.: Microcontroller based electronic distribution board. Int J Sci Eng Res **8**(7) (2017)
5. Tushar, V., Onkar, Y., Ganesh, J., Vishal, D.: Ultra-fast acting electronic circuit breaker for overload protection. In: 3rd International Conference on Advances in Electrical, Electronics, Information Communication and bioinformatics, Chennai, India, 27–28 Feb 2017
6. Zipperer, A., Aloise-Young, P.A., Roche, R., Earle, L., Christensen, D.: Electric energy management in the smart home: perspectives on enabling technologies and consumer behavior. NREL/JA-5500-57586 (2013)
7. Mustafa, G.: Development of a single phase prepaid electrical energy meter using 89S8252 microcontroller architecture. In: 3rd International Conference on Advances in Electrical Engineering, Dhaka, Bangladesh, 17–19 Dec 2015
8. Kravari, K., Kosmanis, T., Papadimopoulos, A.N.: Towards an IOT-enabled Intelligent Energy Management System. 2017 18th International Symposium on Electromagnetic Fields in Mechatronics, Electrical and Electronic Engineering (ISEF) Book of Abstracts, Lodz, Poland, 14–16 Sept 2016
9. Srinivasan, A., Baskaran, K., Yann, G.: IoT based smart plug-load energy conservation and management system. In: 2nd International Conference on Power and Energy Applications, Singapore, 27–30 Apr 2019
10. Patil, N.V., Bondar, D.R., Kanase, R.S., Bamane, P.D.: Intelligent energy meter with advanced billing system and electricity theft detection. In: 2017 International Conference on Data Management (ICDMAI), 24–26 Feb 2017
11. Balamurugan, S., Saravanakamalam, D.: Energy Monitoring and Management Using Internet of Things, Chennai, India, 16–18 Mar 2017
12. Preethi, V., Harish, G.: Design and implementation of smart energy meter. In: Inventive Computation Technologies International Conference, Coimbatore, India, 26–27 Aug 2016
13. Amrapali, D., Kandlikar, W.: Electronic circuit breaker. Int. Res. J. Eng. Technol. (July 2017)
14. Mani, V., G. Abhilasha, Lavanya, Suresh, S.: IOT based smart energy management system. Int. J. Appl. Eng. Res. **12** (2017). ISSN 0973-4562
15. Frolov, V.Y., Bystrov, A.V., Neelov, A.A.: Imitating model of a microprocessor trip unit of a circuit breaker. Young Researchers in Electrical and Electronic Engineering, 2017 IEEE Conference of Russian, St. Petersburg, Russia, 1–3 Feb 2017
16. Machidon, O.M., Stanca, C., Ogrutan, P., Gerigan, C., Aciu, L.: Power system protection device with IoT-based support for integration in smart environment. J. Public Libr. Sci. (2018)
17. Sursum, A.: Residual current circuit breakers. Technical Features and Application Notes, pp. 57–67
18. Ishwar, A.M., Santosh, B.M., Champalal, P.V., J.R, Rokde: Microcontroller based electronics circuit breaker. Int. Res. J. Eng. Technol. (IRJET) **3**(4):569–571 (2016)
19. Himawan, H., Supriyanto, C., Thamrin, A.: Design of prepaid energy meter based on proteus. In: 2nd International Conference on Information Technology, Computer and Electrical Engineering (ICITACEE), Indonesia, 16–18 Oct 2015

Peristaltic Transport of Casson Fluid in a Porous Channel in Presence of Hall Current

M. M. Hasan, M. A. Samad, and M. M. Hossain

Abstract In this present paper, the peristaltic transport of Casson fluid through a porous asymmetric channel has been investigated. Hall current effect is taken into consideration. Mathematical analysis has been considered in a wave frame of reference. The model equations are simplified under the concept of long wavelength and low Reynolds number. Analytic solutions have been obtained for velocity and pressure gradient. The transformed equations have also been solved numerically by bvp4c function from MATLAB. Effects of different involved parameters on velocity and pressure gradient are displayed and explained from the physical point of view. The trapping phenomenon is also discussed. This study reveals that velocity profile increases with an increase in the Hall parameter.

Keywords Hall parameter · Porous channel · Velocity profile · Trapped bolus

1 Introduction

Peristaltic transport is an important mechanism for fluid transport in physiology and industry. This characteristic is naturally associated and occurred with a spontaneous relaxing and compressing movement along the length of the filled channel. This mechanism in channel has a wide range of physiological applications, for examples, urine transport from kidney to bladder, swallowing food material in esophagus, semen movement in the vas deferens of male reproductive tract and blood circulation in small blood vessels [1]. Some examples of peristaltic mechanism in industry are blood pump in heart–lung machine, crude oil refinement, flood processing, sanitary

M. M. Hasan (✉)
Department of Mathematics, Comilla University, Cumilla 3506, Bangladesh
e-mail: marufek@gmail.com

M. M. Hasan · M. A. Samad · M. M. Hossain
Department of Applied Mathematics, University of Dhaka, Dhaka 1000, Bangladesh
e-mail: samad@du.ac.bd

M. M. Hossain
e-mail: mubarakdu@yahoo.com

J. K. Mandal et al. (eds.), *Applications of Internet of Things*, Lecture Notes in Networks and Systems 137, https://doi.org/10.1007/978-981-15-6198-6_3

fluid transport, and noxious fluid transport in nuclear industries. The initial work on peristaltic mechanism in a viscous fluid was conducted by Latham [2], and after that, we found many studies [3–7].

It is mentioned that most of the physiological and industrial fluids are non-Newtonian. We cannot explain all non-Newtonian fluids in one constitutive equation. Thus, a number of non-Newtonian fluid models have been proposed [8]. Casson fluid is one of the non-Newtonian fluids which was introduced by Casson [9]. Human blood can be presented by Casson's model [10]. But little work is done regarding peristaltic transport of Casson fluid with Hall effect. Hall current is essential, and it has noticeable impact on the magnetic force term and current density. The main goal is to study the impact of Hall effect on peristaltic transport of Casson fluid in a porous asymmetric channel. The governing equations are reduced under low Reynolds number and long wavelength approximations. The transformed equations have been solved analytically and numerically. The effects of various important parameters on velocity and pressure gradient are displayed graphically and discussed. Streamline patterns are also sketched. The present study is organized as follows. Section 2 gives the mathematical analysis of the model. Analytic and numerical solutions are determined in Sects. 3 and 4, respectively. The obtained results are explained graphically in Sect. 5. Lastly, the findings of this study are listed in Sect. 6.

2 Mathematical Analysis

A two-dimensional peristaltic transport of a viscous, incompressible, non-Newtonian Casson fluid in a porous and asymmetric channel is considered. Here we choose a stationary frame of reference (X, Y) such that X-axis lies along the direction of channel walls and Y-axis normal to it. Let (U, V) be the velocity components in the stationary frame. The porous medium is assumed to be homogenous. A strong magnetic field with magnitude $\overline{B} = (0, 0, B_0)$ is applied, and the Hall effect is taken into consideration. The induced magnetic field is overlooked for little magnetic Reynolds number. The geometry (in Fig. 1) of the upper and lower wall surfaces is assumed to be

$$\left.\begin{aligned} Y = H_1 = d_1 + a_1 \cos\left\{\tfrac{2\pi}{\lambda}(X - ct)\right\} \\ Y = H_2 = -d_2 - a_2 \cos\left\{\tfrac{2\pi}{\lambda} X - ct + \phi\right\} \end{aligned}\right\} \tag{1}$$

where a_1, a_2 denote the waves amplitudes, $d_1 + d_2$ is the channel width, λ is the wavelength, t is the time, c is the speed of wave propagation, and ϕ is the phase difference changes in the range $0 \leq \phi \leq \pi$. Here, $\phi = 0$ indicates symmetric channel with waves out of phase and $\phi = \pi$ corresponds to waves in phase.

The continuity and momentum equations for incompressible non-Newtonian fluid in absence of body forces are

$$\nabla.\bar{q} = 0 \tag{2}$$

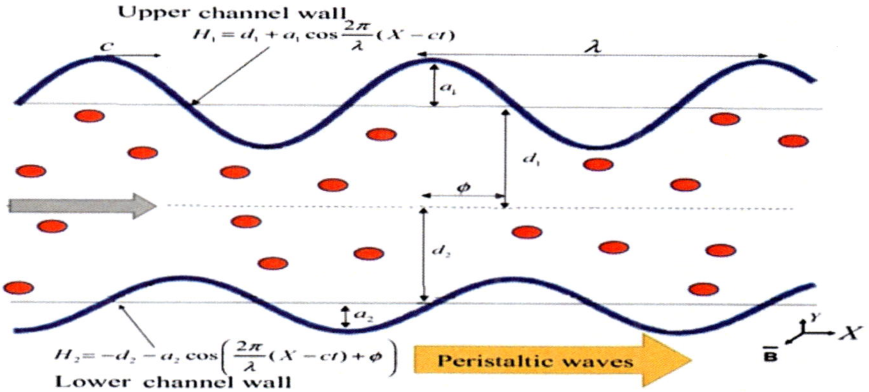

Fig. 1 Geometry of the model

$$\rho \frac{D\bar{q}}{Dt} = -\nabla P + \mu \nabla^2 \bar{q} + \overline{J} \times \overline{B} - \frac{\mu}{K'} \bar{q} \tag{3}$$

Since Hall term is considered, the current density $\overline{J}$ is given by the generalized Ohm's law

$$\overline{J} = \sigma \left[\overline{E} + \bar{q} \times \overline{B} - \gamma \left(\overline{J} \times \overline{B} \right) \right] \tag{4}$$

Using the Maxwell equations, we get

$$\overline{J} \times \overline{B} = \frac{\sigma B_0^2}{1 + m^2} \left[\bar{i}(mV - U) - \bar{j}(V + mU) \right] \tag{5}$$

where $\bar{q}$ is the fluid velocity, P is the pressure, $\gamma = 1/en_e$ is the Hall factor/Hall current, e is the charge electron, n_e is the electron mass, and $\overline{E}$ is the electric field.

The constitute expression for Casson [11] fluid is

$$\tau_{ij} = 2\left(\mu_b + \frac{P_y}{\sqrt{2\pi}} \right) e_{ij} \tag{6}$$

where $e_{ij} = \frac{1}{2}\left(\frac{\partial v_i}{\partial x_j} + \frac{\partial v_j}{\partial y_i} \right)$ is the (i, j) th component of deformation rate, τ_{ij} is the (i, j) th stress tensor component, $\pi = e_{ij}e_{ij}$, and μ_b is the plastic dynamic viscosity. The yield stress P_y is expressed as $P_y = \mu_b \sqrt{2\pi}/\beta$, where β is Casson fluid parameter. For non-Newtonian Casson fluid flow $\mu = \mu_b + P_y/\sqrt{2\pi}$ which gives $\nu' = \nu(1 + 1/\beta)$, where $\nu = \mu_b/\rho$ is the kinematic viscosity for Casson fluid. Again the yield stress $P_y = 0$ for Newtonian case.

Now the governing equations for peristaltic transport of an incompressible Casson fluid through a homogenous porous two-dimensional asymmetric channel are

$$\frac{\partial U}{\partial X}+\frac{\partial V}{\partial Y}=0 \tag{7}$$

$$\frac{\partial U}{\partial t}+U\frac{\partial U}{\partial X}+V\frac{\partial U}{\partial Y}=-\frac{1}{\rho}\frac{\partial P}{\partial X}+\nu\left(1+\frac{1}{\beta}\right)\left(\frac{\partial^2 U}{\partial X^2}+\frac{\partial^2 U}{\partial Y^2}\right) + \frac{\sigma B_0^2}{\rho\left(1+m^2\right)}(mV-U)-\nu\left(1+\frac{1}{\beta}\right)\frac{U}{K'} \tag{8}$$

$$\frac{\partial V}{\partial t}+U\frac{\partial V}{\partial X}+V\frac{\partial V}{\partial Y}=-\frac{1}{\rho}\frac{\partial P}{\partial Y}+\nu\left(1+\frac{1}{\beta}\right)\left(\frac{\partial^2 V}{\partial X^2}+\frac{\partial^2 V}{\partial Y^2}\right) - \frac{\sigma B_0^2}{\rho\left(1+m^2\right)}(mU+V)-\nu\left(1+\frac{1}{\beta}\right)\frac{V}{K'} \tag{9}$$

The boundary conditions are

$$\left.\begin{array}{l} U=0 \text{ when } Y=H_1 \\ U=0 \text{ when } Y=H_2 \end{array}\right\} \tag{10}$$

where B_0 is the uniform magnetic field strength, σ is the electric conductivity, ρ is the fluid density, and K' is the permeability of the porous space.

The flow is not steady in the stationary frame (X, Y), but it turns into steady in the moving wave frame (x, y). The stationary frame and wave frame are linked to

$$x=X-ct,\ y=Y,\ u=U-c,\ v=V,\ p(x,y)=P(X,Y,t) \tag{11}$$

Here u, v, p are the velocity components and pressure in the (x, y) frame, respectively. To reduce the difficulty of the model equations, we use the following dimensionless quantities.

$$\left.\begin{array}{l} x'=\frac{x}{\lambda},\ y'=\frac{y}{d_1},\ u'=\frac{u}{c},\ v'=\frac{v}{c\delta},\ t'=\frac{ct}{\lambda},\ p'=\frac{pd_1^2}{\lambda c\mu_b} \\ \delta=\frac{d_1}{\lambda},\ h_1=\frac{H_1}{d_1},\ h_2=\frac{H_2}{d_1},\ d=\frac{d_2}{d_1},\ a=\frac{a_1}{d_1},\ b=\frac{a_2}{d_1} \end{array}\right\} \tag{12}$$

The governing Eqs. (7)–(9) under the assumptions of long wavelength and low Reynolds number in terms of stream function ψ (dropping the das symbols) become

$$\frac{\partial p}{\partial x}=\left(1+\frac{1}{\beta}\right)\left[\frac{\partial^3\psi}{\partial y^3}-\alpha^2\left(\frac{\partial\psi}{\partial y}+1\right)\right] \tag{13}$$

$$\frac{\partial^4\psi}{\partial y^4}-\alpha^2\frac{\partial^2\psi}{\partial y^2}=0 \tag{14}$$

$$\frac{\partial p}{\partial y}=0 \tag{15}$$

The reduced boundary conditions are

$$\left.\begin{array}{l}\psi = \frac{q}{2},\quad \frac{\partial \psi}{\partial y} = -1 \quad \text{when } y = h_1 = 1 + a\cos 2\pi x \\ \psi = \frac{-q}{2},\quad \frac{\partial \psi}{\partial y} = -1 \text{ when } y = h_2 = -d - b\cos(2\pi x + \phi)\end{array}\right\} \tag{16}$$

where $\alpha^2 = \frac{M^2}{(1+m^2)(1+1/\beta)} + \frac{1}{K}$, $M = \sqrt{\frac{\sigma}{\mu_b}} B_0 d_1$ is the magnetic field parameter, $K = K'/d_1^2$ is the permeability parameter, q is the mean flow rate in the wave frame, $m = \sigma\gamma B_0$ is the Hall parameter, and $\beta = \frac{\mu_b\sqrt{2\pi}}{P_y}$ is the Casson fluid parameter.

Note that Q and q be the dimensionless forms of mean flow rate in stationary frame and wave frame, respectively. Also they are related by

$$Q = q + 1 + d \tag{17}$$

in which

$$q = \int_{h_2}^{h_1} u \mathrm{d}y \tag{18}$$

3 Analytic Solution

Exact solutions of reduced governing equations along with the boundary conditions (16) were obtained by direct integration. The solutions of stream function ψ, velocity u, and pressure gradient $\frac{\mathrm{d}p}{\mathrm{d}x}$ are

$$\begin{aligned}\psi &= A_1 + A_2 y + A_3\cosh(\alpha y) + A_4\sinh(\alpha y)\\ u &= A_2 + A_3\alpha\sinh(\alpha y) + A_4\alpha\cosh(\alpha y)\\ \frac{\mathrm{d}p}{\mathrm{d}x} &= -(1 + 1/\beta)\alpha^2(A_2 + 1)\end{aligned}$$

where

$$\begin{aligned}A_1 &= \frac{-(h_1 + h_2)\left[q\alpha\cosh\frac{\alpha}{2}(h_1 - h_2) + 2\sinh\frac{\alpha}{2}(h_1 - h_2)\right]}{2(h_1 - h_2)\alpha\cosh\frac{\alpha}{2}(h_1 - h_2) - 4\sinh\frac{\alpha}{2}(h_1 - h_2)}\\ A_2 &= \frac{q\alpha\cosh\frac{\alpha}{2}(h_1 - h_2) + 2\sinh\frac{\alpha}{2}(h_1 - h_2)}{(h_1 - h_2)\alpha\cosh\frac{\alpha}{2}(h_1 - h_2) - 2\sinh\frac{\alpha}{2}(h_1 - h_2)}\\ A_3 &= \frac{(h_1 - h_2 + q)\sinh\frac{\alpha}{2}(h_1 + h_2)}{(h_1 - h_2)\alpha\cosh\frac{\alpha}{2}(h_1 - h_2) - 2\sinh\frac{\alpha}{2}(h_1 - h_2)}\end{aligned}$$

$$A_4 = \frac{-(h_1 - h_2 + q)\cosh\frac{\alpha}{2}(h_1 + h_2)}{(h_1 - h_2)\alpha\cosh\frac{\alpha}{2}(h_1 - h_2) - 2\sinh\frac{\alpha}{2}(h_1 - h_2)}$$

4 Numerical Solution

The transformed equations have also been solved numerically for different values of model parameters using MATLAB software (bvp4c function). Also the following data has been used: $a = 0.5$, $b = 0.4$, $d = 1.5$, $q = -1$, $x = 0$, $M = 1$, $K = 0.5$, $\beta = 0.5$, $m = 1$, $\phi = \pi/4$, unless otherwise specified. The value of Prandtl number for human blood is $Pr = 21$ [6]. So Pr is kept 21 in this study. Numerical computations have been carried out for various values of magnetic field parameter (M), Casson fluid parameter (β), permeability parameter (K), flow rate (q), and Hall parameter (m). The software ORIGIN has been used to show the numerical results graphically.

5 Results and Discussions

The behavior of magnetic field parameter M on velocity component u is plotted in Fig. 2. It is clear that when M is increased, the fluid velocity u diminishes. The reason is that applied magnetic field produces a resistive force to the flow and this force diminishes the velocity of the fluid. The magnitude of the velocity u increases with increasing Hall parameter m as seen in Fig. 3. The fact is that the effective conductivity $\sigma/(1 + m^2)$, existed in the momentum equation, diminishes when m increased which consequently reduces the magnetic Lorentz force. Therefore, velocity increases. Figure 4 is sketched to see the variation of velocity profiles

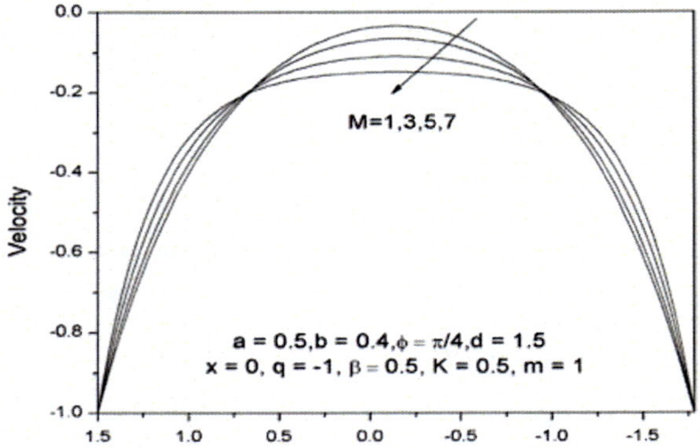

Fig. 2 Velocity profiles for M

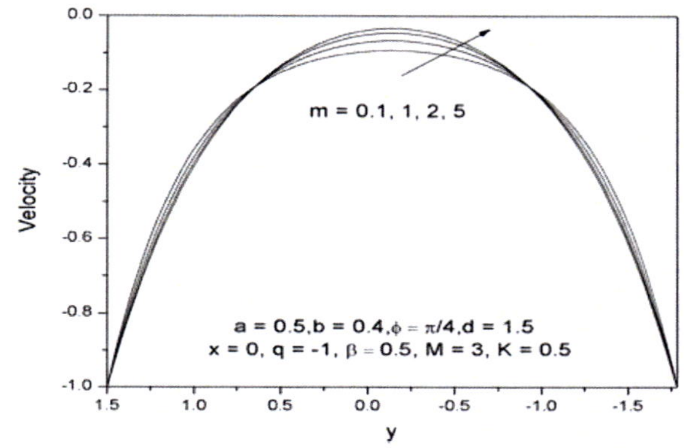

Fig. 3 Velocity profiles for *m*

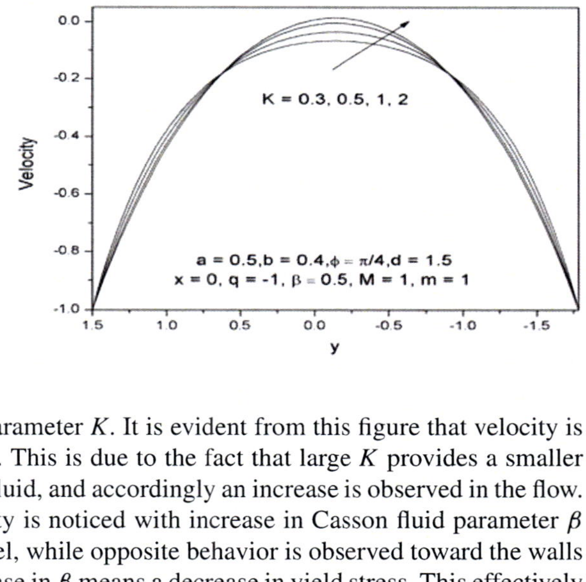

Fig. 4 Velocity profiles for *K*

for different permeability parameter K. It is evident from this figure that velocity is an increasing function of K. This is due to the fact that large K provides a smaller amount of resistance to the fluid, and accordingly an increase is observed in the flow. Again an increase in velocity is noticed with increase in Casson fluid parameter β near the center of the channel, while opposite behavior is observed toward the walls as shown in Fig. 5. An increase in β means a decrease in yield stress. This effectively accelerates the fluid flow. Figure 6 shows that the velocity profile increased with an increase in flow rate q.

The effects of M, m, K, β and q on pressure gradient $\mathrm{d}p/\mathrm{d}x$ over one wave length $x \in [0, 1]$ are shown in Figs. 7, 8, 9, 10, and 11. Figure 7 shows $\mathrm{d}p/\mathrm{d}x$ increases for large M. On the other hand, pressure gradient $\mathrm{d}p/\mathrm{d}x$ reduces for increasing m, K, β and q.

Another interesting phenomenon is trapping for peristaltic flow. It depends on the formulation of contours of streamlines. The impacts of magnetic field parameter M and Hall parameter m on streamline patterns are shown in Figs. 12 and 13. It is noted that the volume of the entangled bolus reduces with an increase in M. Magnetic field parameter M is the ratio of magnetic force to inertia force. Increase in M enhances a

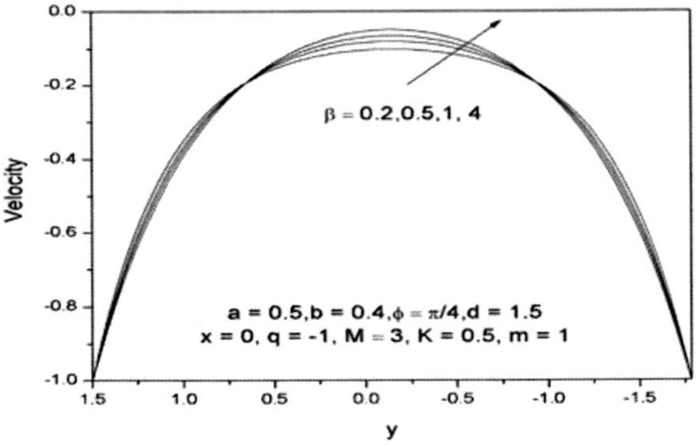

Fig. 5 Velocity profiles for β

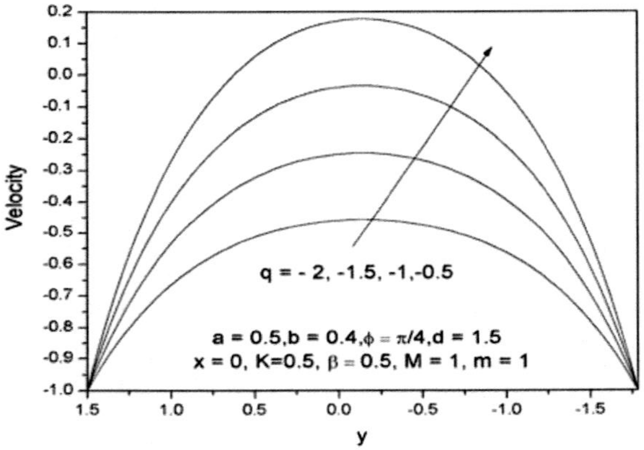

Fig. 6 Velocity profiles for q

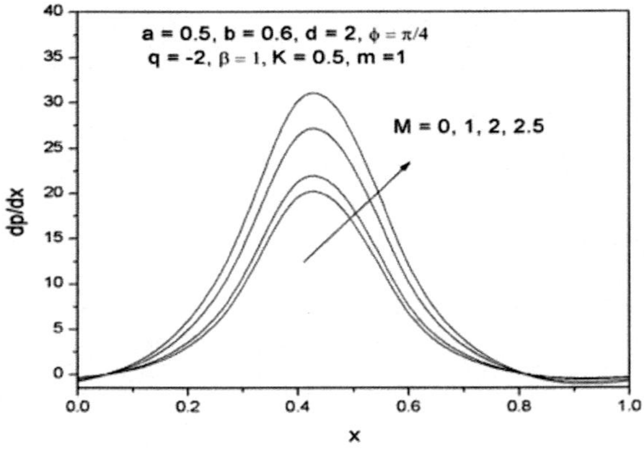

Fig. 7 Pressure gradient for M

Fig. 8 Pressure gradient for m

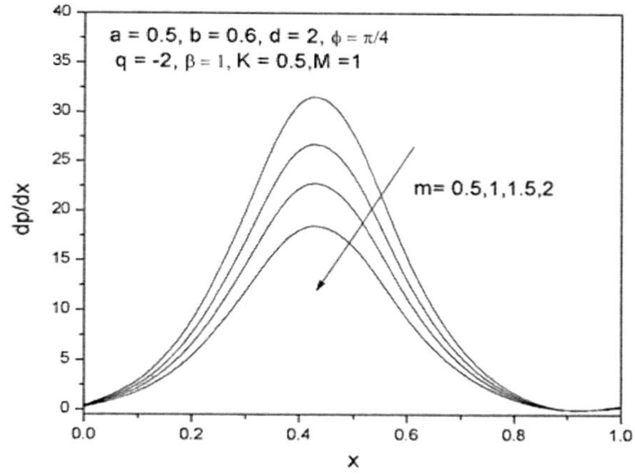

Fig. 9 Pressure gradient for K

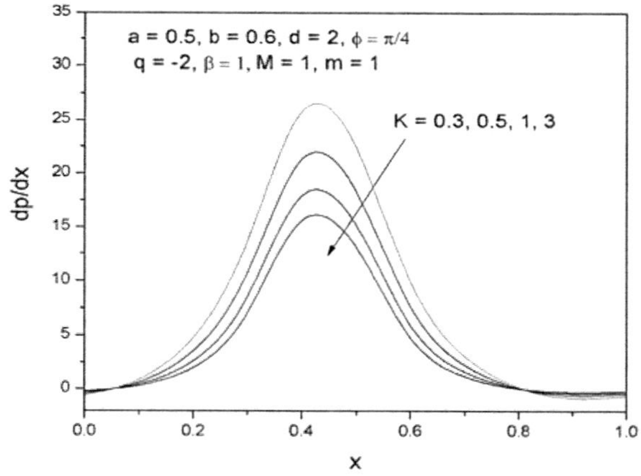

Fig. 10 Pressure gradient β

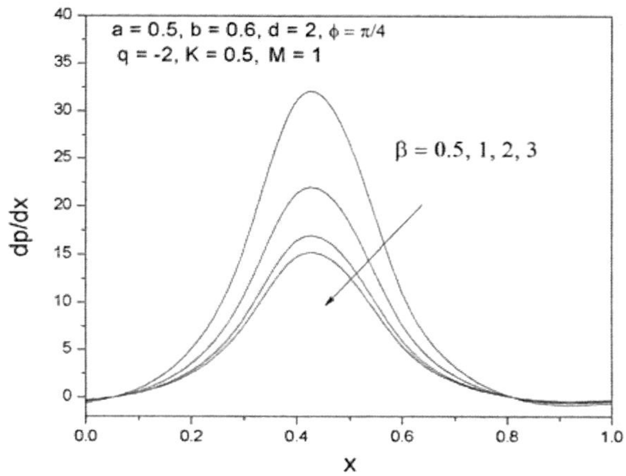

Fig. 11 Pressure gradient for q

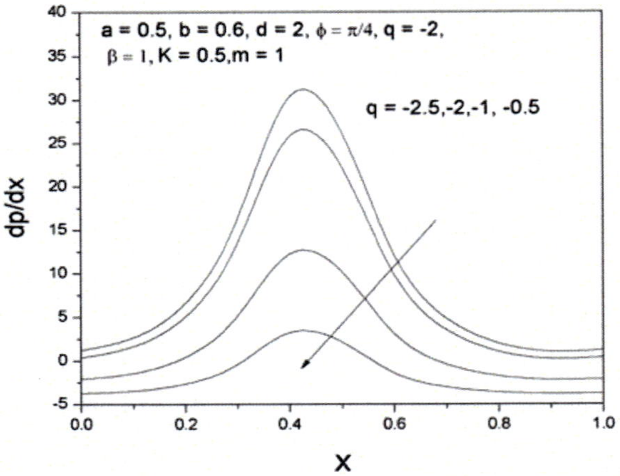

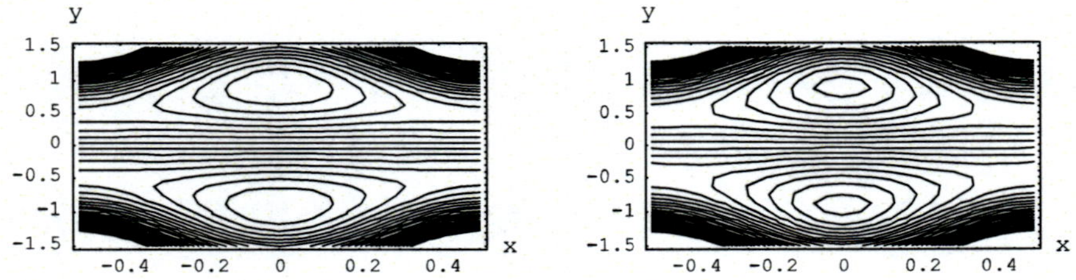

Fig. 12 Streamline patterns for (**a**) $M = 1$ and (**b**) $M = 3$

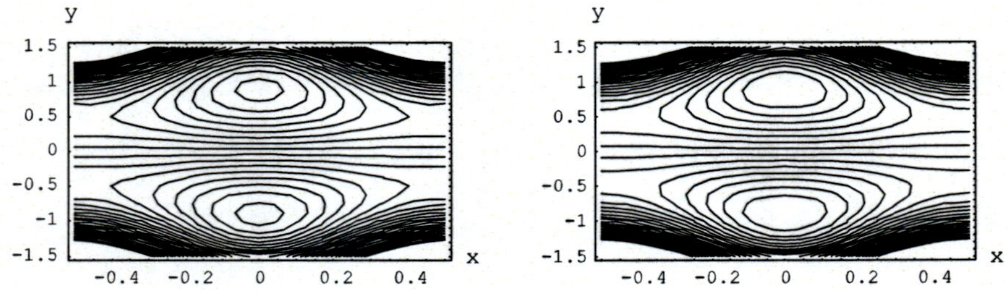

Fig. 13 Streamline patterns for (**a**) $m = 0.5$ and (**b**) $m = 1$

force, and this force causes the resistance in the flow of the fluid. Again the size of the trapped bolus magnifies with an increase in m as shown in Fig. 13.

Figure 14 gives the comparison between the results obtained in the present study and the results of previous study [12]. To do so, both the studies have been brought to the same platform, by considering equal parameter values (Newtonian case).

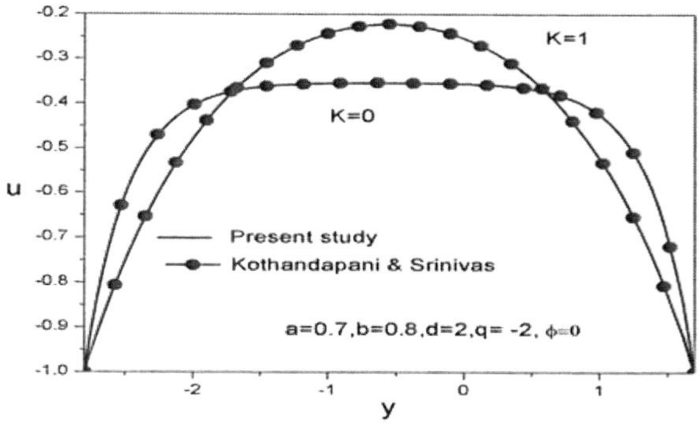

Fig. 14 Comparison of velocity profiles

6 Conclusion

This study is presented on the peristaltic transport of Casson fluid in a porous channel. Hall current effect was taken into consideration. The main findings of the study are as follows:

1. Hall parameter has an increasing impact on velocity profile.
2. Pressure gradient decreases for K, m, β and q.
3. The size of the trapped bolus enlarges for m.

References

1. Yildirim, A., Sezer, S.A.: Effects of partial slip on the peristaltic flow of a MHD Newtonian fluid in an asymmetric channel. Math. Comput. Model. **52**, 618–625 (2010)
2. Latham, T.W.: Fluid motion in a peristaltic pump, MS thesis, MIT Cambridge, MA (1966)
3. Akbar, N.S., Butt, A.W.: Physiological transportation of Casson fluid in a plumb duct. Commun. Theor. Phys. **63**(3), 347–352 (2015)
4. Ahmed, B., Javed, T., Ali, N.: Numerical study at moderate Reynolds number of peristaltic flow of micropolar fluid through a porous-saturated channel in magnetic field. AIP Adv. **8**(1), 015319-1-16 (2018)
5. Hayat, T., Ali, N.: Peristaltically induced motion of MHD third grade fluid in a deformable tube. Phys. Lett. A **370**, 225–239 (2006)
6. Misra, J.C., Sinha, A.: Effect of thermal radiation on MHD flow of blood and heat transfer in a permeable capillary in stretching motion. Heat Mass Transfer **49**, 617–628 (2013)
7. Hasan, M.M., Samad, M.A., Hossain, M.M.: Peristaltic flow of non-newtonian fluid with slip effects: analytic and numerical solutions. Res. J. Math. Stat. **11**(1), 1–10 (2019)
8. Nadeem, S., Ul Haq, R., Lee, C.: MHD flow of a Casson fluid over an exponentially shrinking sheet. Sci. Iranica **19**(6), 1550–1553 (2012)
9. Casson, N.: Rheology of Disperse Systems. Pergamon Press, London, p. 84 (1959)
10. Blair, S., William, G., Spanner, D.C.: Introduction to biorheology by GW Scott Blair, Chapter XII on Botanical Aspects by DC Spanner (1974)

11. Eldabe, N.T.M., Saddeck, G., El-Sayed, A.F.: Heat transfer of MHD non-Newtonian Casson fluid flow between two rotating cylinders. Mechan. Mech. Eng. **5**(2), 237–251 (2001)
12. Kothandapani, M., Srinivas, S.: Non-linear peristaltic transport of a Newtonian fluid in an inclined asymmetric channel through a porous medium. Phys. Lett. A **372**, 1265–1276 (2008)

Fingerprint Authentication System for BaaS Protocol

Ranadhir Debnath, Swarup Nandi, and Swanirbhar Majumder

Abstract Over the past many years, several corporations have benefited from the implementation of cloud solutions among the organization. Due to the advantages such as flexibility, mobility, and cost saving, the number of cloud users is expected to grow rapidly. Consequently, organizations want a secure system, credit to manifest its users so as to make sure the practicality of their services and information hold on within the cloud storages are managed in a private environment. In the current approaches, the user authentication in cloud computing is predicated on the credentials submitted by the user like secret, token and digital certificate. Unfortunately, these credentials can often be stolen, accidentally revealed, or hard to remember. In view of this, we propose a fingerprint-based authentication system to support the user authentication for the cloud environment. We take into account a distributed state of affairs wherever the biometric templates are hold on within the cloud storage, whereas the user authentication is performed without the leak of any sensitive information.

Keywords Biometric authentication · Fingerprint recognition · BaaS protocol · Minutiae

1 Introduction

Biometrics is measure of biological or behavioral features which are used for identification of individuals. Most of these features are inherited and cannot be guessed or stolen [1]. Biometric systems are based on two techniques, physiological (face, voice, fingerprint, iris, etc.) and behavioral (signature, etc.). Biometric characteristics

R. Debnath · S. Nandi (✉) · S. Majumder
Department of Information Technology, Tripura University, Suryamaninagar, Agartala 799022, Tripura, India
e-mail: swarupnandi@tripurauniv.in

R. Debnath
e-mail: ranadhirdebnath02@gmail.com

S. Majumder
e-mail: swanirbhar@ieee.org

J. K. Mandal et al. (eds.), *Applications of Internet of Things*, Lecture Notes in Networks and Systems 137, https://doi.org/10.1007/978-981-15-6198-6_4

such as iris patterns, face, fingerprints, palm prints, and voice will be submitted by the user as the credential for authentication over the cloud. Biometric-based authentication systems provide a higher degree of security as compared with conventional authentication systems [2].

A fingerprint is an impression left by the friction ridges of a human finger [3]. Particular inspects show that no two individuals have comparable fingerprints, so they are unique for each person [4]. Fingerprint is an important feature which can uniquely identity an individual. Fingerprint of two different persons can never be same. Because of this unique feature, fingerprints are extremely well known for biometric authentication applications. There are edges and valleys in human unique fingerprints. When they are combined, they shape particular examples which get grew a short time later and are called unique fingerprints [5]. Each fingerprint can be identified by minutiae which are some uncommon parts on edges. Further minutiae are divided into two sections: termination and bifurcation. Termination is called as completion, and bifurcation is called as branch [6]. Fingerprints are used for uncommon distinguishing proof or acknowledgment by individual during the long period [7]. Present-day fingerprint matching techniques were begun in the early sixteenth century [8]. A basic development and progress in unique fingerprint identification were made in 1899 by Edward Henry, perceived as the popular "Henry system" of fingerprint classification [7, 8]: a detailed technique for ordering fingerprints especially tuned to encouraging and helping the human specialists [7, 8] shown in Fig. 1.

BaaS is Biometric-as-a-Service. Banking service is an end-to-end process ensuring the overall execution of a financial service provided over the web. Such

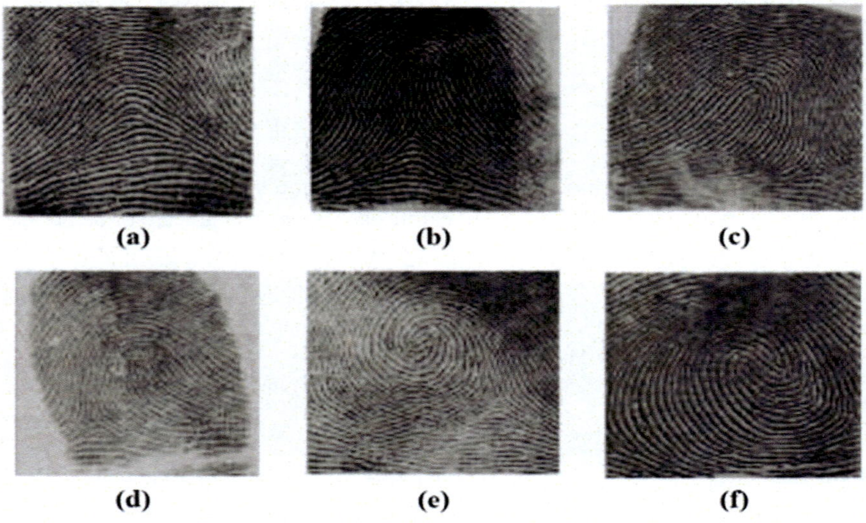

Fig. 1 Fingerprints and a fingerprint classification schema involving six categories. **a** arch, **b** tented arch, **c** right loop, **d** left loop, **e** whorl, and **f** twin loop

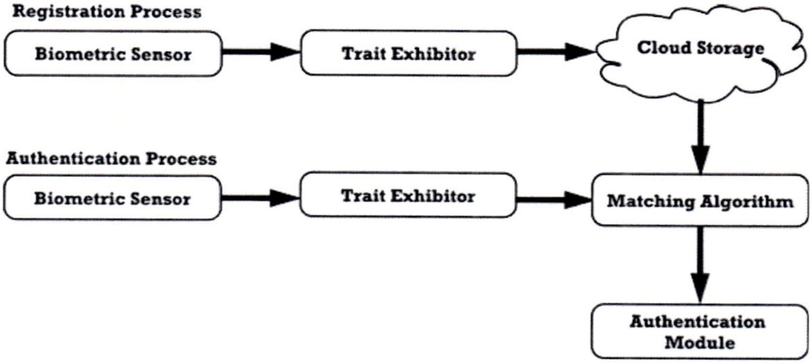

Fig. 2 Block diagram of BaaS protocol-based authentication system

a digital banking service is available on demand and is carried out within a set time-frame [9]. BaaS (Biometric-as-a-Service) framework performs biometric matching in cloud operations. This framework normally relies on popular consumer devices like smartphones with simple fingerprint sensors [10]. Biometric-as-a-Service (BaaS) provides single sign-in for user verification or authentication. Fingerprint authentication system based on BaaS protocol has lots of advantages over existing conventional authentication system [10]. When user's secret data are being considered, user validation in the cloud environment is necessary and it is done by using BaaS protocol. In Baas, matching algorithm is required for verification or validation of user [11] shown in Fig. 2. Biometrics-as-a-service (BaaS) is a model that uses the well-dug in practices of the SaaS model (Software-as-a-Service) that performs biometric matching tasks in the cloud environment and gives it as a service [12] (Fig. 3).

So typical fingerprint authentication system consists of five components:

1. **Image capture**: In this component, a sensor captures fingerprint data in digital format.
2. **Preprocessing**: In this component, the input image is improved using various image processing techniques such as histogram equalization, fast Fourier transform.
3. **Feature extraction**: After improvement, the minutiae, which are ridges and valleys of a unique fingerprint, are extracted.
4. **Template generation**: A template is created consists of extracted minutiae. In case of enrollment process, the template is stored in the template database and

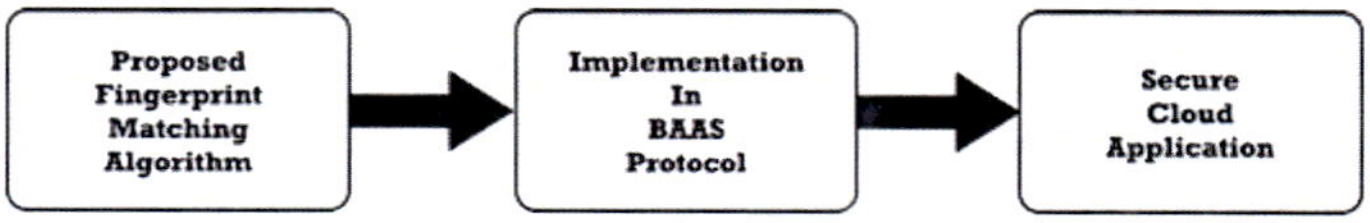

Fig. 3 Block diagram of fingerprint authentication system implemented in BaaS protocol

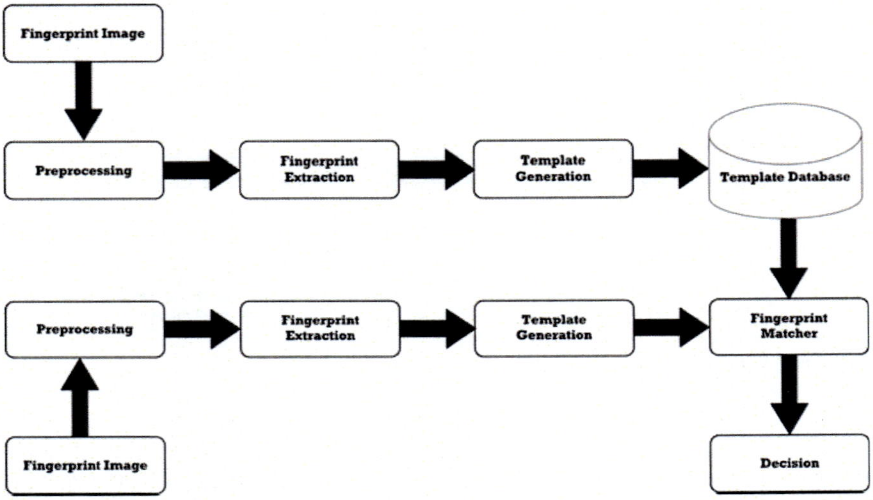

Fig. 4 Block diagram of fingerprint authentication system

in case of authentication process, the template is sent to the next component for matching.

5. **Fingerprint matching**: The received template is matched with the templates stored in the template database, and decision (fingerprint verified or not) is made (Fig. 4).

Srivastava et al. [13] proposed a fingerprint matching algorithm consists of three stages, viz. prehandling stage, minutiae extraction stage, post-handling stage, and prehandling stage steps are histogram equalization and fast Fourier transform, which improves the image; and then, binarization and segmentation are done on the improved fingerprint image. Minutiae extraction stage has two phases: edge diminishing and minutiae checking. Post-handling stage has only one step to remove the fake minutiae.

Sagayam et al. [14] proposed a new fingerprint authentication algorithm which uses Euclidean distance and artificial neural network (ANN). Preprocessing (histogram equalization and fast Fourier transform) is done on the input image. Then, binarization and thinning are done on the processed image. After that minutiae are extracted. Euclidean distance classification is done on training set and testing set, and the performance is analyzed using NN classifier.

Almajmaie et al. [15] proposed a fingerprint recognition system based on modified multi-connect architecture (MMCA). The algorithm steps include preprocessing step, recognition step, and identification. Segmentation and image binarization are done in preprocessing step on the input image. MMCA is applied on the recognition step. MMCA is done in two phases, training and analysis phases.

2 Proposed Fingerprint Authentication Algorithm

Our proposed fingerprint matching algorithm for BaaS protocol consists of following steps:

1. Input image preprocessing
   a. Scaling
   b. Masking
   c. Histogram equalization
   d. Parallel and orthogonal smoothing
   e. Binarization
2. Minutiae extraction and template creation
   a. Create skeleton map of Minutiae
   b. Masking inner Minutiae
   c. Minutiae cloud removal
   d. Extracting top Minutiae
   e. Shuffling of Minutiae.
3. Fingerprint matching.

Image Preprocessing

The input image is scaled to 500 dpi. If the image is more or less than 500 dpi, else no scaling is done as the input image is of 500 dpi. Every single other part of the calculation expects they work with 500dpi pictures. Filtered mask is applied on the scaled image by using some set of filters to mark the valid fingerprint area. The histogram equalization is performed on the scaled image. After histogram equalization, two separate images are created, one applying parallel smoothening and another applying orthogonal smoothing from equalized image. Binarized image is a version of fingerprint image that has all pixels set either to black or white with no shades of gray. It is computed during preprocessing by comparing parallel smoothing of the image to orthogonal smoothing.

Minutiae Extraction and Template Creation

First step in Minutiae extraction after image binarization is to create skeleton map of minutiae which contains ridges and valleys of the fingerprint image. The skeleton which has only one connected ridge is considered. Endings are skeleton minutiae from ridges skeleton, and bifurcations are valley skeleton. Inner minutiae are masked using some set of filters. A step called Minutiae cloud removal is done to skeleton minutiae after inner minutiae masking to remove the minutiae clouds which are tightly packed constellations of minutiae that are errors in early steps of the algorithm. Finally, the template is obtained, which is top minutiae skeleton map from skeleton map after removing minutiae clouds. Top minutiae eliminate all the low-quality minutiae to form a high-quality skeleton map. Then, shuffled minutiae is applied which is the last filter. This does not remove any minutiae; it just changes the minutiae order randomly. The random ordering is reliable, which implies that running feature extractor on a similar image twice results in the very same unique fingerprint template.

Fingerprint Matching

Edge table is a list of the nearest neighbor details for each reference minutiae in fingerprint template is created. It is calculated from shuffled minutiae during minutiae extraction process. For each pair of reference and neighbor minutiae, edge table contains an edge that recognizes neighbor minutiae (reference minutiae are verifiable from table structure) and portrays edge shape: edge length, relative reference, and relative neighbor bearing. Edge shape is a translation-invariant and rotation-invariant finger-print feature suitable for matching. Then, edge tables of both images are compared to generate a score, and then authentication is done comparing the score with the predefined threshold value. If the score is more than the threshold value, then both the images match, else the images does not match.

3 Results and Discussion

To check the effectiveness of our proposed fingerprint matching algorithm, we tested our algorithm using FVC (2004) database. The database contains four folders (DB1, DB2, DB3, and DB4). Each folder contains 80 different fingerprint images and 8 impressions of each fingerprint. The images are grayscale images with resolution of 500 dpi (http://bias.csr.unibo.it/fvc2004).

Some images of FVC 2004 database are shown in Fig. 5.

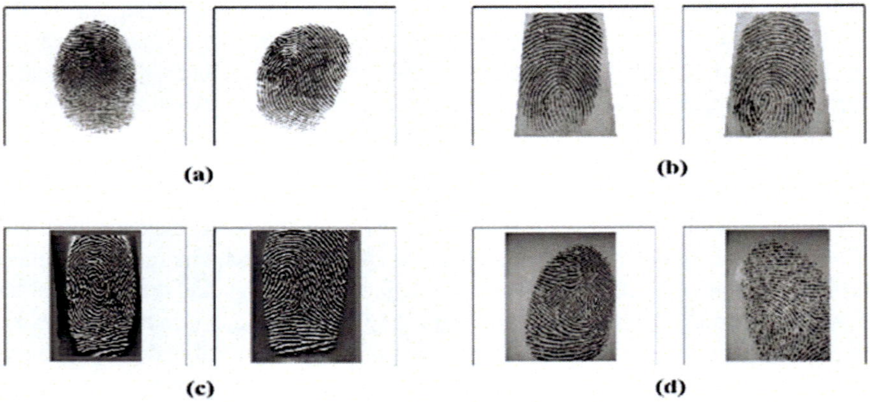

Fig. 5 Some fingerprints of FVC 2004 database. **a** DB1 folder, **b** DB2 folder, **c** DB3 folder, and **d** DB4 folder

The accuracy of the fingerprint authentication algorithm is measured using the confusion matrix and determining five parameters, viz. accuracy, true positive rate, true negative rate, precision, and retail.

Terminologies related to the test:

1. **True positive (TP)**: Both the comparing images are of same person, and result is positive, i.e., the fingerprint matches.
2. **False positive (FP)**: Both the comparing images are of different person, and result is positive, i.e., the fingerprint matches.
3. **True negative (TN)**: Both the comparing images are of same person, and result is negative, i.e., the fingerprint does not match.
4. **False negative (FN)**: Both the comparing images are of different person, and result is negative, i.e., the fingerprint does not match.

Formulae for measuring the parameters:

$$\text{Accuracy} = \frac{\text{TP} + \text{TN}}{\text{TP} + \text{TN} + \text{FP} + \text{FN}}$$

$$\text{True Positive Rate} = \frac{\text{TP}}{\text{TP} + \text{FN}}$$

$$\text{True Negative Rate} = \frac{\text{TN}}{\text{TN} + \text{FP}}$$

$$\text{Precision} = \frac{\text{TP}}{\text{TP} + \text{FP}}$$

$$\text{Recall} = \frac{\text{TP}}{\text{TP} + \text{FN}}$$

From Fig. 6, in all the cases we get 100% TN rate and 0% FN rate, which we expected, but this is not the same for TP and FP which we expected as 100% and 0%, respectively, but we could not achieve. The highest TP is 84.72% when sample size is 6, while lowest TP is 79.3% when sample size is 8, and the average TP is 81.99% which is almost 82%. And for FN which average stands at 18%, got highest value of 20.71% with sample size 8 and lowest with 15.28% when sample size is 6. Thus, we can say when we take sample size 6, our system gives more accurate results than any other sample size and less accurate when size is 8.

The average confusion matrix according to Fig. 6 is:

From Fig. 7, it is found that the FAR, which is false accept rate of our system is 0% and FRR, which is false reject rate is 18%.

From Fig. 8, we can find that our algorithm is almost 91% accurate and we get highest accurate results when sample size is 6.

A line graph of accuracy, TPR (true positive rate), TNR (true negative rate), precision, and recall is shown in Fig. 4. In the graph, TPR and recall are represented in the same straightline as both are 100%.

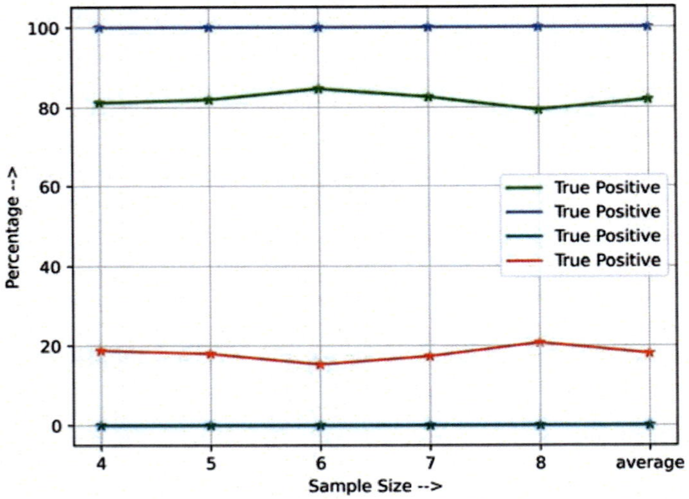

Fig. 6 Comparison of TP, FP, TN, FN by taking various image samples

Fig. 7 Average confusion matrix

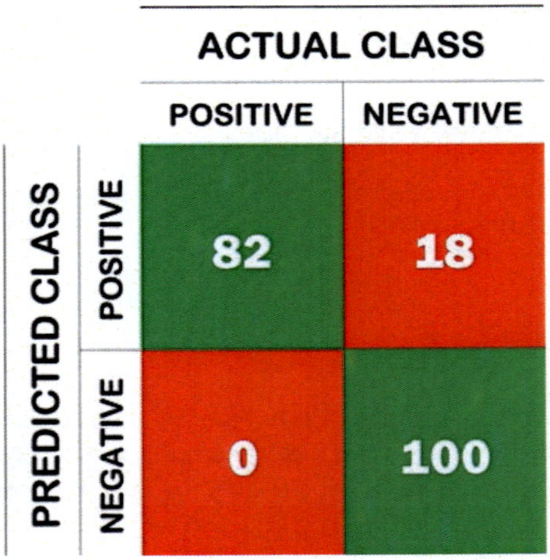

4 Conclusion

In this work, a false accept rate (FAR) of 100% was obtained, but it had 18% false reject rate (FRR), which means 18 out of 100 times, the system will reject authentication when it is supposed to be authenticated. The algorithm needs to be modified in future to reduce FRR as well as implement it on the BaaS protocol in cloud. The implementation in cloud with the fingerprint biometric that shall facilitate higher

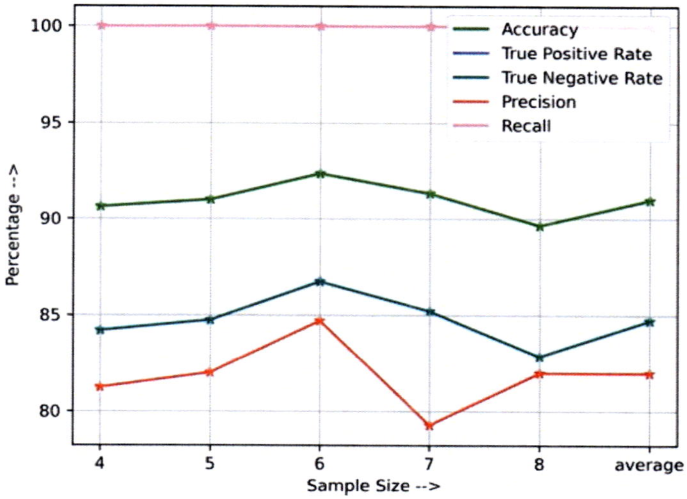

Fig. 8 Graphical comparison of accuracy, TPR, TNR, precision, and recall

security along with liveness detection shall help in cases of intellectual properties and applications like money or revenue transaction.

References

1. What is Biometrics? https://www.geeksforgeeks.org/what-is-biometrics/. Last accessed on 8 Jan 2020
2. Wong, K.-S., Kim, M. H.: Secure biometric-based authentication for cloud computing. In: Ivanov et al. (Eds.): CLOSER 2012, CCIS 367, pp. 86–101 (2013)
3. Fingerprint-Wikipedia. https://en.wikipedia.org/wiki/Fingerprint. Last accessed on 8 Jan 2020
4. Barham, Z. S., Mousa, A.: Fingerprint recognition using MATLAB. Bachelor Diss (2011)
5. Tatsat Naik, O.S.: Fingerprint Recognition System, pp. 141–144. Springer, New York (2003)
6. Nallaperumall, K., Fred, A.L., Padmapriya, S.: A novel for fingerprint feature extraction using fixed size templates. In: IEEE2005 Conference, pp. 371–374 (2005)
7. Gaw, A.: Lee and Gaensslen's Advances in Fingerprint Technology. CRC Press (2012)
8. Federal Bureau of Investigation, United States: The Science of Fingerprints: Classification and Uses. US Department of Justice, Federal Bureau of Investigation (1984)
9. Banking Service: Wikipedia. https://en.wikipedia.org/wiki/Banking_service. Last accessed on 08 Jan 2020
10. Swarup, N., Majumder, S.: Overview of liveliness detection of fingerprint for using it in BaaS protocol in cloud. Int. J. Comput. Intell. IoT **2**(4) (2018)
11. Sepasian, M., Mares, C., Balachandran, W.: Vitality detection in fingerprint identification. Inf. Sci. Appl. **4** (2010)
12. Mantra Blog: What is biometrics-as-a-service—Mantra Blog. https://blog.mantratec.com/what-is-biometric-as-a-service. Last accessed on 08 Jan 2020
13. Srivastava, A. P., et al. Fingerprint recognition system using MATLAB. In: 2019 International Conference on Automation, Computational and Technology Management (ICACTM). IEEE (2019)

14. Sagayam, K.M., et al.: Authentication of biometric system using fingerprint recognition with euclidean distance and neural network classifier. Int. J. Innov. Technol. Explor. Eng. **8**(4), 766–771 (2019)
15. Almajmaie, L., Ucan, O.N., Bayat, O.: Fingerprint Recognition System Based on Modified Multi-Connect Architecture (MMCA). Cognitive Systems Research (2019)

Design of a Low-Cost Li-Fi System Using Table Lamp

Suman Debnath and Bishanka Brata Bhowmik

Abstract This paper presents a designing of a Li-Fi working model to send information in a unidirectional path via visible light to a receiving device across free space. The communication link will be set up between a mobile device and a PC using a modified table lamp to transmit data serially via USB COM port.

Keywords Light fidelity (Li-Fi) · Visible light communication (VLC) · Radiofrequency (RF) · Universal asynchronous receiver/transmitter (UART) · COM (communication) port

1 Introduction

A rapid evolution in technology is not only helping the society to progress, but it also opens the door of a new era of creative thinking for future innovations. Li-Fi is one such emerging technology in the subset of visible light communication (VLC) where the data communication is done wirelessly by modulating the output intensity of the light-emitting diodes (LEDs) with respect to the binary information, whereas a photo-detector is used at the receiver end to recover the transmitted signal.

Li-Fi was coined by a German professor Harald Hass that stands for *Light Fidelity*. He demonstrated this concept of optical wireless communication (OWC) at the TED Global Talk in Edinburgh in 2011 [1]. The concept of using light as medium of transmission dates back to the ancient times when light is being used in various forms like smoke signals or beacon fires to convey messages [2]. Over the years, optical communication has been evolved to a more advanced form where data nowadays is being sent wirelessly via optical medium that proved to be a complementary technology to the existing radio-frequency (RF) communication [3]. Li-Fi uses license-free visible

S. Debnath (✉) · B. B. Bhowmik
Department of Electronics and Communication Engineering, Tripura University, Suryamaninagar, Agartala 799022, Tripura, India
e-mail: debnathsuman91@gmail.com

B. B. Bhowmik
e-mail: bishankabhowmik@tripurauniv.in

J. K. Mandal et al. (eds.), *Applications of Internet of Things*, Lecture Notes in Networks and Systems 137, https://doi.org/10.1007/978-981-15-6198-6_5

light spectrum (375–780 nm) to provide a short-range wireless link for data communication. The concept was first proposed by the Japanese researchers in the form of VLC. It was in the year 2000 a group of researchers from Japan proposed and simulated successfully the concept using a LED-based indoor wireless transmitting station [4]. From then on, this field attracts a lot of attention across the globe.

Till date, a few start-up companies are offering products based on this technology. Among them, PureLi-Fi [5], Ledcomm [6], Velmenni [7], etc., are prominent who tested and came up with some good solutions for practical approach to implement the technology. PureLi-Fi introduced the Li-Fi-XC a USB dongle capable for full bi-directional multiuser communication via light. Currently, they are working on various components like Gigabit Li-Fi and Li-Fi ASIC [7]. Li-Fi MAX, GEOLi-Fi OEM modem, etc. products are offered by Ledcomm.

This paper demonstrates a working model of a light-based communication link between two devices via serial port. A detailed explanation of a Li-Fi transmitter along with the receiver has been shown.

2 Working Principle

Li-Fi is a type of visible light communication (VLC) that works on the principle of modulating a light source to convey information which is detected by a photo-detector and processing circuitry stationed at the receiving end to recover the original information [8]. Low-cost low–power-consuming LEDs are used as the light source that gives very bright luminescence modulated by switching it on and off with the help of a driver circuit at a high frequency [9].

The modulation of the LEDs is carried out by various modulation techniques. If the modulation is done based on the technique such that the LEDs remains on if the binary bit '1' and turns off for binary bit '0', then it is working on OOK (on-off keying) modulation format. It is a widely used single-carrier modulation (SCM) scheme for its easy implementation [10]. In comparison to SCM, multicarrier modulation (MCM) schemes are used for high-speed multiuser applications. MCM schemes are more efficient in terms of energy and bandwidth. A widely used MCM technique known as orthogonal frequency division multiplexing (OFDM) can also be used to transmit data streams simultaneously in parallel with the help of different orthogonal subcarrier.

The transmitted data that is passed through the optical medium falls on the sensitive area of the optical detector circuitry. The circuitry consists of a photo-sensitive element or sensor to detect the modulated light signal. The sensor converts the light in the form of current proportional to it, and hence, the light gets detected at the receiving end. Depending upon the modulation used at the transmitting side, the receiving circuitry is designed that can demodulate the receiving signal to the original data [11]. Generally, photo-sensitive element like a light-dependent resistor (LDR) or a photo-diode or a photo-transistor can be used to detect the incoming light signal.

After the detection, the signal is feed to a transimpedance amplifier circuitry before demodulation to recover the information.

3 Design of a Li-Fi System

In this section, a detailed explanation of the working model of the Li-Fi system is presented. The model consists of a transmitter and a receiver circuitry.

3.1 Transmitter Circuitry

Li-Fi transmitter converts the digital data into visible light. For the light source, white high-brightness LEDs were used. The transmitter modulates the LEDs on the basis of the incoming data to be sent. The modulation format used here is the OOK modulation. Based on this format, the circuit turns on the LEDs to transmit logic one and it turns off the LEDs to transmit logic zero.

The data transmission is done via serial port, so a serial device is used. Figure 1 shows the designed transmitter. The serial device is connected to the COM port of the transmitting device via USB. The connected device is a Silicon Labs CP2102 USB to TTL UART converter. The output TX pin of the serial converter is feed to the base pin of a switching transistor (2N2222A) that drives the SMD LEDs.

The LEDs are connected to the 5 V optional output power pin of the TTL converter. In this way for an incoming bit high or low, the variation of the TX pin output will change the state of the transistor to turn on and off the LEDs.

3.2 Receiver Circuitry

The receiver circuit detects the incoming light signal, amplifies, and compares it to get the desired output.

Figure 2 shows the receiver circuitry. A low-cost light-dependent resistor (LDR) device is used to detect the light signal which is connected to achieve a potential divider circuit. The potential divider output is then feed to the non-inverting terminal of the dual op-amp LM358 IC, while a 10 K potentiometer is connected to the inverting terminal of the same op-amp IC. Thus, the op-amp works as a comparator that compares and amplifies the voltage difference of the two input terminals to produce the output.

A LED is connected across the output terminal of the op-amp to indicate the output sequence. The output of the op-amp 1 is feed to the op-amp 2 that acts as a buffer circuit, and the final output is obtained from the op-amp 2.

The circuit diagram of the receiver circuitry is shown in Fig. 3. The distance

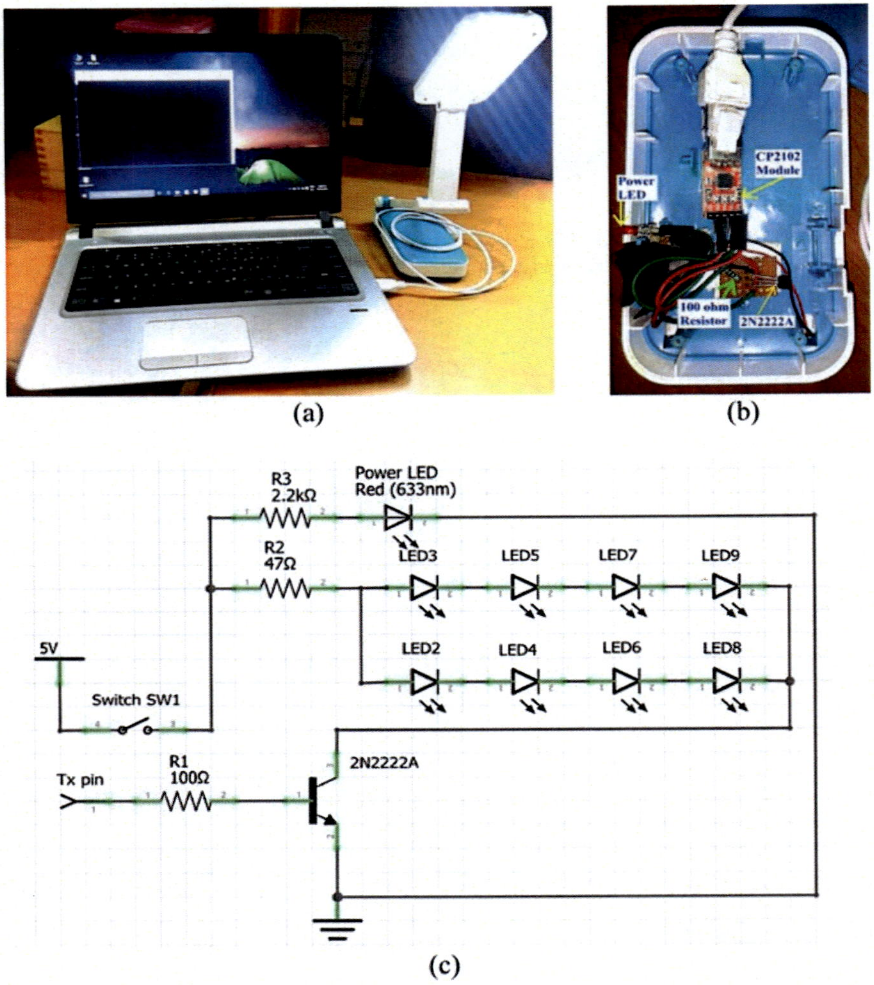

Fig. 1 **a** Transmitter unit connected to PC, **b** transmitter internal construction, and **c** transmitter circuit diagram

between the light source and the LDR can be adjusted with the help of the potentiometer. A CP2102 USB to TTL UART converter is used in which the output is connected to the RX pin to convert the incoming bits back to the USB standard. The converter is then connected to the USB port of the receiving device where it will be detected as a specific COM port device.

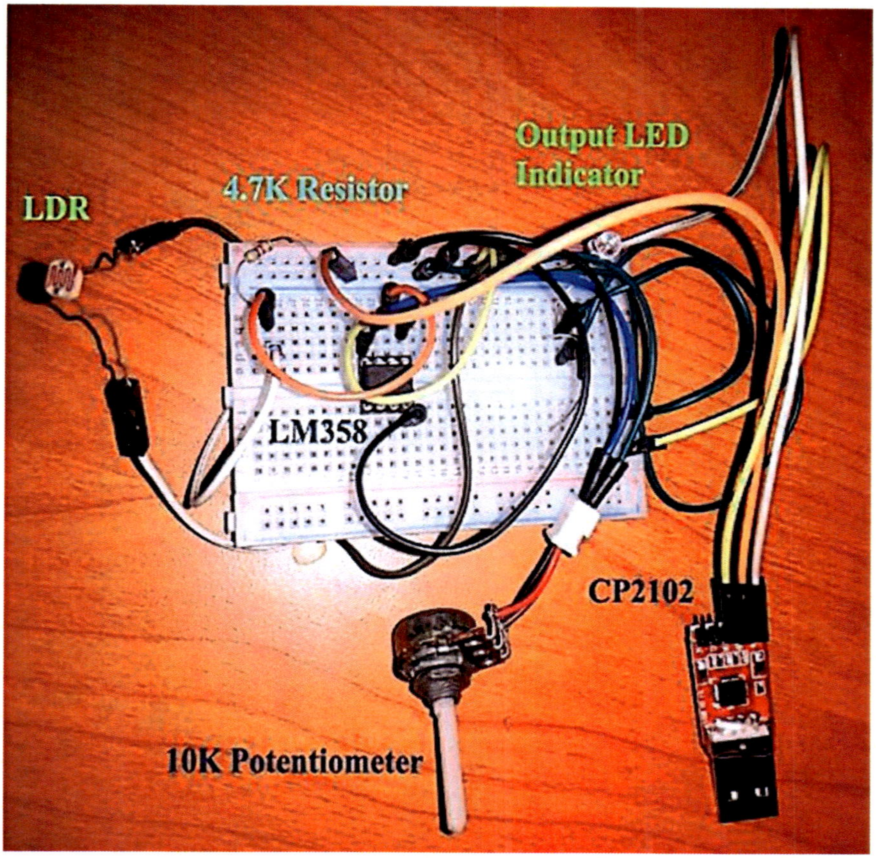

Fig. 2 Receiver circuitry

3.3 Software

The communication link has been set up between a mobile and a PC device using visible light. For this, open-source software like *Serial USB Terminal* and *Tera Term* have been used for demonstrating the transfer of text contents between these devices via serial port. The software automatically detects the transmitter and the receiver connected to the COM port. After setting up the connection between the COM ports with the software, the serial port has been manually configured to adjust the baud rate, data, parity, and stop bits.

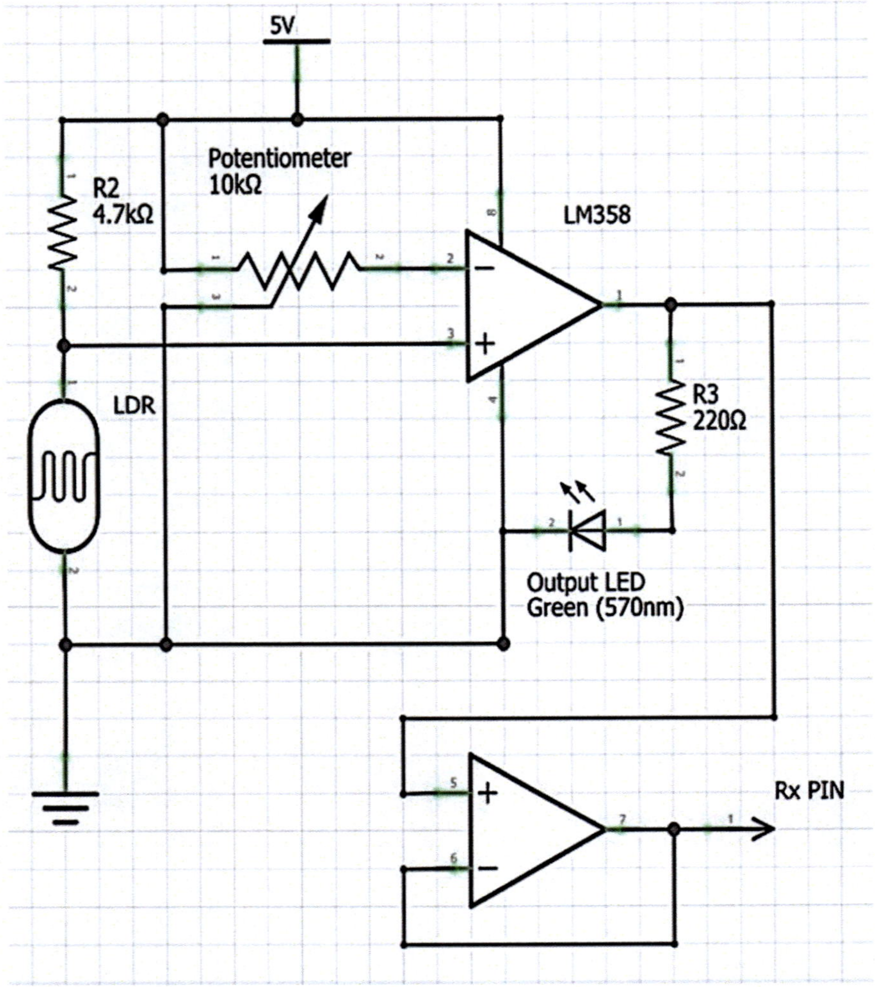

Fig. 3 Receiver circuit diagram

4 Results

The transmitter and receiver results of the Li-Fi communication link are shown in Fig. 4.

Figure 4 shows the waveforms of the transmitted and the received signals. The received signal is obtained after amplifying the sensor output. Though the waveforms obtained are almost identical in nature, there exists a small difference in phase and duty cycle between them. This shows that the data transmission is feasible.

The figures of the serial terminals are shown in Fig. 5. A string of data is trans-

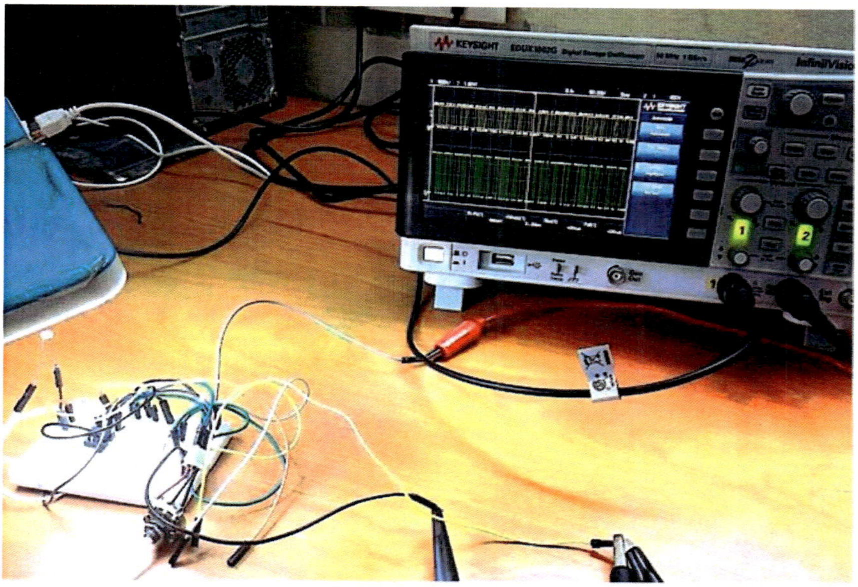

(a)

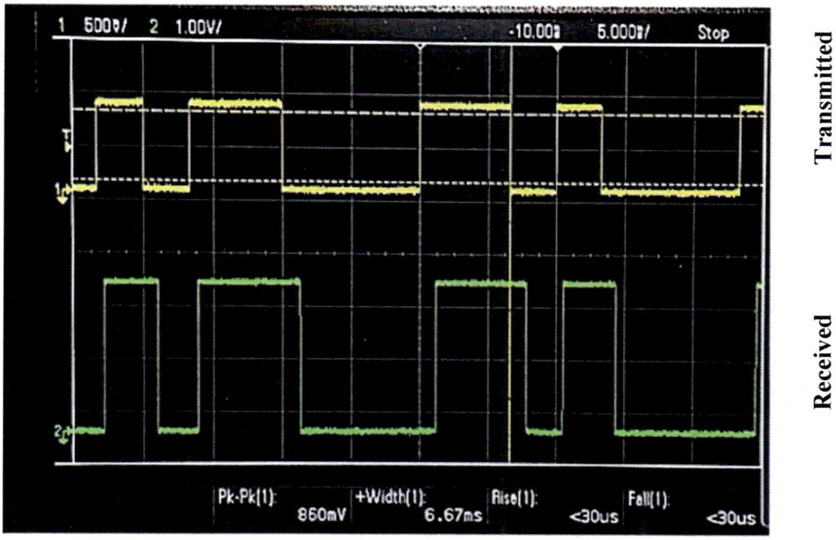

(b)

Fig. 4 **a** Hardware setup for testing in DSO, and **b** transmitted and received signals obtained

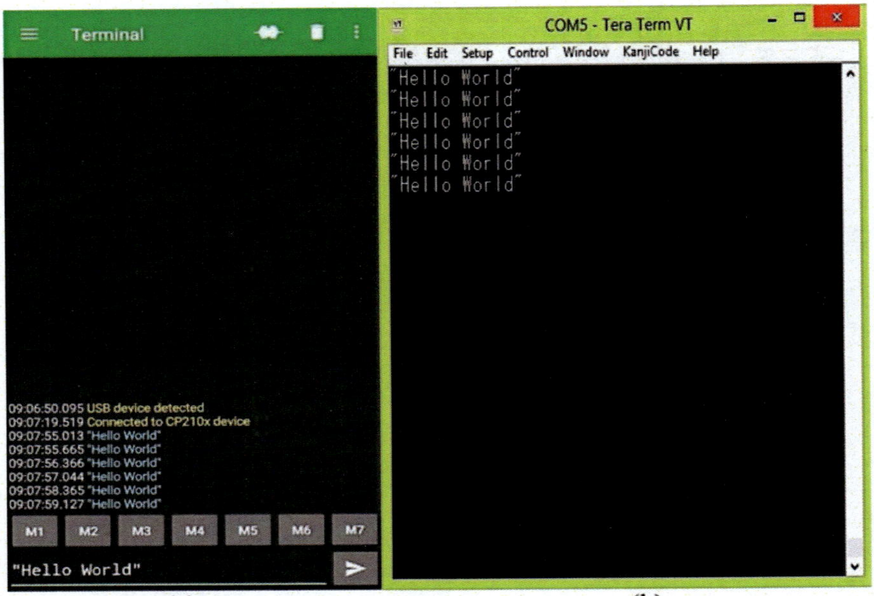

Fig. 5 **a** Data string transmitted from mobile to PC and **b** received data string on PC

mitted from the mobile to pc using *Serial USB Terminal* application with the help of the designed transmitter. The data is received at the receiving end and finally been displayed at the *Tera Term* terminal monitor screen. A saved text file can also be send via this setup.

5 Conclusion

In this paper, a working model of a Li-Fi-based communication link has been successfully demonstrated. The transmitter and receiver model has been presented in detail. The model has been used to transmit and receive data strings over visible light using LEDs. The communication is done by modulating the light intensity using on-off keying technique. From the experimented demonstration, it is shown that the feasibility of data transmission using visible light is possible. The model designed has some limitations also like speed, accessibility, and direction of propagation. The design does not support multiuser bi-directional access. Further, if the receiver is not placed at a required distance and also not in the line of sight of the transmitter, the data transmission gets affected. Though the paper aims to present an easy, compact, and low-cost Li-Fi communication link, the speed and distance between the transmitter and the receiver can be further increased with the help of high-speed devices.

References

1. The History of LiFi. https://lifi.co/the-history-of-lifi/. Last accessed 29 Sept 2019
2. Dimitrov, S., Haas, H.: Principles of LED Light Communications towards Networked Li-Fi, 1st edn. Cambridge University Press, Cambridge (2015)
3. Bian, R., Tavakkolnia, I., Haas, H.: 15.73 Gb/s Visible light communication with off-the-shelf LEDs. J. Lightwave Technol. 1 (2019). https://doi.org/10.1109/jlt.2019.2906464
4. Nan, Chi: LED-Based Visible Light Communications. Tsinghua University Press, Springer, Beijing, Germany (2018)
5. PureLiFi. https://purelifi.com/. Last accessed 25 Sept 2019
6. Oledcomm. https://www.oledcomm.net/. Last accessed 26 Sept 2019
7. Velmenni. https://www.velmenni.com/. Last accessed 28 Sept 2019
8. https://purelifi.com/lifi-products/. Last accessed 25 Sept 2019
9. Shamsudheen, P., Sureshkumar, E., Chunkath, Job: Performance analysis of visible light communication system for free space optical communication link. Proc. Technol. **24**, 827–833 (2016). https://doi.org/10.1016/j.protcy.2016.05.116
10. Haas, H., Yin, L., Wang, Y., Chen, C.: What is Li-Fi? J. Light Wave Technol. **34**, 1533–1544 (2016)
11. Goswami, P., Shukla, M.K.: Design of a Li-Fi transceiver. Wirel. Eng. Technol. **8**, 71–86 (2017). https://doi.org/10.4236/wet.2017.84006

A Study of Micro-ring Resonator-Based Optical Sensor

Papiya Debbarma, Srikanta Das, and Bishanka Brata Bhowmik

Abstract Optical ring resonator evolved as a latest technology in recent years for various sensing applications. This paper focused refractive index-based sensing capabilities of ring resonator in optical light detection explained the ring resonator sensors designs and reviews the present state of the field. Several factors have been taken into account during simulation, including the effect of ring radius, gap spacing, input wavelength, refractive index, and waveguide width and height.

Keywords Ring resonator · Optical sensor · Refractive index-based sensor

1 Introduction

Optical ring resonator consists of waveguides; among these minimum one is a closed loop which is attached to some kind of light input and output [1]. To understood how optical ring resonator work, we must understand the optical path length (L_{optical}) of a ring resonator. This is given for a single-ring resonator.

$$\text{OPD} = 2 * \text{pi} * r * n_{\text{eff}} \tag{1}$$

Here, r = Radius of the ring.

n_{eff} = effective index of refractive in waveguide material.

A sensor may be defined as a device, component, or subsystem whose function is to detect actions or changes in its environment and send the information to other electronics, frequently a computer processor. Optical sensor has long been popular

P. Debbarma (✉) · S. Das · B. B. Bhowmik
Tripura University, Suryamaninagar, Agartala 799022, Tripura, India
e-mail: papiyadb96@gmail.com

S. Das
e-mail: Srikanta.ece@tripurauniv.in

B. B. Bhowmik
e-mail: bishankabhowmik@tripurauniv.in

J. K. Mandal et al. (eds.), *Applications of Internet of Things*, Lecture Notes in Networks and Systems 137, https://doi.org/10.1007/978-981-15-6198-6_6

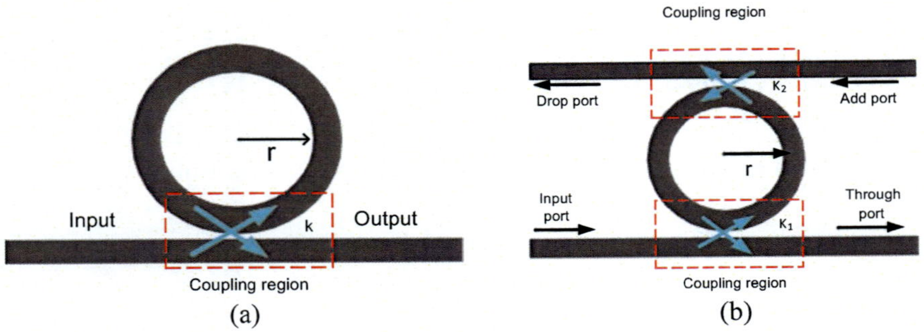

Fig. 1 Basic model of an optical ring resonator. **a** Ring configuration of a one straight waveguide (all pass structure), **b** two straight waveguide coupled to each other through a ring waveguide (add-drop structure). k = coupling efficient, r = ring radius

for analysis of a various type's gas or liquid. It is a high-sensitivity, optical sensor also allows rapid analysis, high specificity due to specific light matter interaction, low interaction, with samples. An optical sensor changes over light beams into an electronic sign. The motivation behind an optical sensor is to quantify a physical amount of light and, contingent upon the sort of sensor, at that point makes an interpretation of it into a structure that is intelligible by an incorporated estimating gadget. Optical sensors are utilized for contactless recognition, checking or situating of parts. Optical sensors can be either inward or outward.

There are different types of sensors:

- Electrical sensor
- Mechanical sensor
- Optical sensor
- Chemical sensor
- Thermal sensor.

An optical sensor is basically used for detecting light intensity. It converts the light ray into an electronic signal, measures the physical quantity of light, and converts to a readable form to an instrument. Advantages of optical sensors:

- Completely passive (can be used in an explosive environment).
- Resistance to high temperature and pressure and also chemically reactive environment.
- High sensitivity, bandwidth range, and better resolution.
- Lightweight and small size.

2 Literature Survey

The basic structure of ring resonator was first proposed in the year of 1969 by E. A. J. Marcatili. In that paper, he proposed designs having round or rectangular cross section and straight axis [2].

In 1989, C. Casper and E. J. Bachns have studied about a kind resonator which helped into construction and calculation the manufacturing and measurement of a fiber micro-ring resonator of wavelength 1.55 μm with a bi-conical reduced fiber diameter of 8.5 μm. The diameter 2-mm ring exhibits a large free spectral range of 30 GHz. The power coupling coefficient was measured to be 0.28, and the insertion loss is 1.2 dB with a finesse of 4.7, from the optical frequency response [3].

In 1991 was taken by P. A. Bernard and J. M. Gautray done an experiment to measure the permittivity of dielectric medium with the micro-strip ring resonator. This concept was introduced in relation to the calculation done of the line capacitance of a multilayer micro-strip so that it can effectively make arrangements of effective permittivity and resonant frequency of the ring. The result of ring resonator was compared with the measurements made in X-band waveguide cavity by cavity perturbation technique. In this experiment first, the measurements were taken using the ring resonator removed between space tool devices. In circulator reflected power is obtained at port 3. That is means the result tends to assure that the micro-strip ring resonator can be used for measurement of dielectric [4].

The micro-ring resonator also worked as channel dropping filters. In 1995, a paper has been published by B. E. Little, T. Chu, H. A. Hans; they proposed side coupled to an optical signal bus in micro-ring and disk resonator. The erbium amplifier bandwidth to enclose with abundant free spectral range served as a channel dropping filters [5].

In 1997, B. E. Little, S. T. Chu, H. A. Hans along with J. Foresi and J. P. Laine described a micro-ring resonator with a linear waveguide coupled to a ring waveguide. In this experiment, multiple rings were coupled to obtain high-order filters which resulted with improved pass-band characteristics. It has also been observed that multiple coupled rings can make a huge difference in filter performance by providing layer out of band signal rejection [6].

In 2010, Yuze Sun, Xudong Fan came up with an idea to generate optical ring resonators to create biochemical and chemical sensing technology so as to detect analytics in liquid or gas. A ring resonator sensing principle was therefore introduced which defined different ring resonator sensor designs. Researches are done these days specially to detect samples from more complex media. Emphasis is given more to act on the objective to rely on sensitive label-free ring resonator sensing technology. So that we can replace the currently used fluorescence-based detection, for example enzyme-linked immunosorbent assay [1].

Ricardo Marchetti et al. during 2017 represent the optimized micro-ring resonator, the standard ones SOI depending on si-waveguide with lower height (less than 220 nm), constructed by silicon-on-insulator (SOI) field, using approved lithographic optical filter with an insulator loss, lower than 1 dB empower the comprehension of high-ability optical filter [7].

In 2014, Lo S. M. et al. showed light interaction with matter. They have worked on the project to detect the sensitivity of PhCR consisting biosensors which due to huge refractive index changes. Specific DNA and protein were also tested. In this case, it has been found that the bulk refractive index is ~248\RIU, which is more than that the ordinary micro-ring devices. Through biosensing of DNA and protein at the nanomolar level, PhCRs are known to have more than twofold surface to enhance detection sensing than the used micro-ring resonator devices [8].

In 2017, Nihal F. studied and monitor a highly noble and sensitive photonic crystal (PhC) refractometer for glucose concentration. In the proposed design was based on a two-dimensional photonics crystal platform with face-shaped defect field with the analyzed analytics. Performance of the biosensor was investigated; the structural geometric parameter and sensitivity is maximum [9].

3 Micro-ring Resonator

A micro-ring resonator is an advanced optical component with an affluence of applications especially in the fields of switching, routing, and sensing. Since 1990, it has become one of the widely used optical components in the field of integrated optics technology. It is used to enclose light by total internal reflection, which can be generated by micro- or nanofabrication techniques. A ring resonator is usually comprised of a straight waveguide along with a circular one.

The light passes through the linear waveguide, and it is combined with circular waveguide by the transient field. Resonator sensitivity is determined by the shift of the resonance wavelength; it happen due to the change in the refractive index of the sampled type [10].

$$\Delta\lambda_{\text{res}} = (\Delta n_{\text{eff}} L/m) \quad m = 1, 2, 3 \tag{2}$$

where m = resonance mode.

L = circumference of the ring waveguide.

Where there is no sharp edge in the object, the full-wave half maximum (FWHM) is widely used. It is a 3 dB resonance width free spectral range (FSR) which refers the distance between two consecutive fringes of the resonator, and it is defined as:

$$\text{FSR} = \lambda^2/2\pi.ng.R \tag{3}$$

The ratio between FSR and resonance width is called finesses. The finesse is determined by:

$$\text{Finesse} = \lambda_{\text{res}}/\text{FWHM} \tag{4}$$

When light passes through waveguide, it covers the total no of circular distance which defines the sharpness relative to the central frequency in that ring waveguide is known as Q-factor [11].

$$\text{Q-factor} = \lambda_{res}/\text{FWHM} \tag{5}$$

4 Optical Coupling

It is primarily the spillage of light from one straight waveguide to the another. In the ring resonator, if the straight waveguide and the ring waveguide are close to enough to one another, some portion of the light in the linear waveguide spilled into the ring because of wave property of light, transmission impact. Optical coupling light basically depends on the three different characteristics like the length between linear and ring waveguide, the coupling length, and the refractive index between the linear and ring waveguide.

As the distance becoming closer, the coupling should be straightforward and superior. The curve length is the integrated length of a ring resonator when it is close to the linear waveguide. It will be easy to couple, if the coupling length increases [12–16].

In the simulation, the construction of a silicon-based ring resonator is shown. For simulation purpose, FDTD was used. The model basically consists of ring resonator waveguide, a linear waveguide. The ring simulation has been done in the Opti-FDTD software. Figure 3 shows the electromagnetic field of steady state of the wave which reaches the steady-state condition. The lightwave that travels inside the ring waveguide and the lightwave inside the linear waveguide interferes with each other and hence create a resonance spectrum (Fig. 4).

Figure 5 shows the change in refractive index in different resonance shift. In every 1 nm, refractive index changes 0.02%

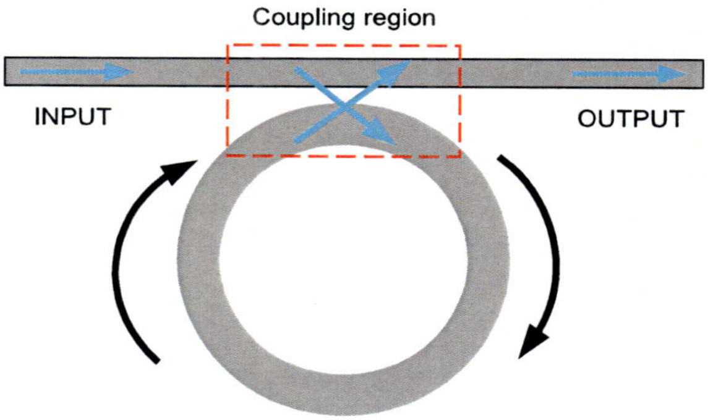

Fig. 2 Micro-ring resonator with coupling region

Fig. 3 Steady-state distribution of electromagnetic field

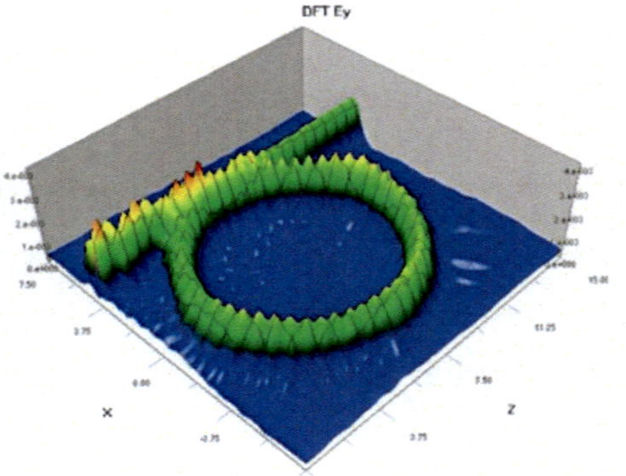

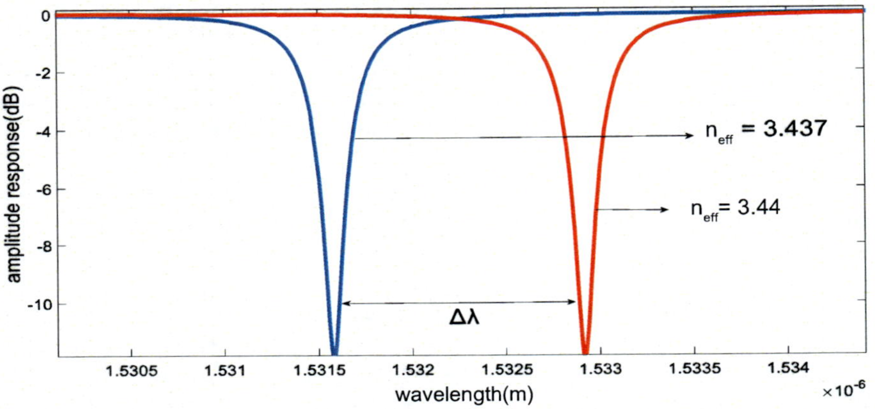

Fig. 4 Shift in resonance wavelength of the ring resonator with respect to change in refractive index

Fig. 5 Sensitivity curve

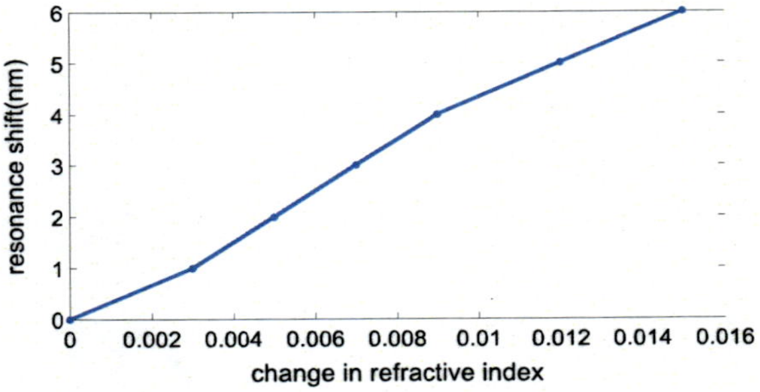

5 Conclusion

In this paper, we have presented a study on all-pass micro-ring resonator which is a very promising element to be an optical sensor. The vital parameters like ring resonator, optical sensor, and refractive index-based sensor of a micro-ring resonator have been discussed; along with that the basic working principle of refractive index-based micro-ring resonator-based sensor has been stretched in this study.

References

1. Sun, Y., Fan, X.: Optical ring resonators for biochemical and chemical sensing. Anal. Bioanal. Chem. **399**(1), 205–211 (2011)
2. Marcatili, E.A.J.: Bends in optical dielectric guides. Bell Syst. Tech. J. **48**(7), 2103–2132 (1969)
3. Caspar, C., Bachus, E.J.: Fibre-optic micro-ring-resonator with 2 mm diameter. Electron. Lett. **25**(22), 1506–1508 (1989)
4. Bernard, P.A., Gautray, J.M.: Measurement of dielectric constant using a microstrip ring resonator. IEEE Trans. Microw. Theory Tech. **39**(3), 592–595 (1991)
5. Little, B.E., Chu, S.T., Haus, H.A.: Micro-ring resonator channel dropping filters. In: LEOS'95. IEEE Lasers and Electro-Optics Society 1995 Annual Meeting. 8th Annual Meeting. Conference Proceedings, vol. 2, pp. 233–234. IEEE (1995)
6. Little, B.E., Chu, S.T., Haus, H.A., Foresi, J., Laine, J.P.: Microring resonator channel dropping filters. J. Lightwave Technol. **15**(6), 998–1005 (1997)
7. Marchetti, R., Vitali, V., Lacava, C., Cristiani, I., Giuliani, G., Muffato, V., Fournier, M., Abrate, S., Gaudino, R., Temporiti, E., Carroll, L.: Low-loss micro-resonator filters fabricated in silicon by CMOS-compatible lithographic techniques: design and characterization. Appl. Sci. **7**(2), 174 (2017)
8. Lo, S.M., Hu, S., Gaur, G., Kostoulas, Y., Weiss, S.M., Fauchet, P.M.: Photonic crystal microring resonator for label-free biosensing. Opt. Express **25**(6), 7046–7054 (2017)
9. Areed, N.F., Hameed, M.F.O., Obayya, S.S.A.: Highly sensitive face-shaped label-free photonic crystal refractometer for glucose concentration monitoring. Opt. Quant. Electron. **49**(1), 5 (2017)
10. Rifat, A.A., Ahmed, R., Bhowmik, B.B.: SOI waveguide-based biochemical sensors. In: Computational Photonic Sensors, pp. 423–448. Springer, Cham (2019)
11. Das, N., Brata, B.: A study on microring resonator based sensor in the health sector. Int. J. Comput. Intell. IoT **2**(4) (2019)
12. Carloni, A.: Ntp Nano Tech Projects Srl. Laser optical coupling for nanoparticles detection. U.S. Patent 10,133,048 (2018)
13. Takayama, S., Abe, K., Fujii, R., Honda, T., Chen, S.X., Harakawa, O., SAE Magnetics (HK) Ltd.: Coupling structure of optical components and coupling method of the same. U.S. Patent 10,061,084 (2018)
14. Li, D., Chang, W., Liu, C., Liu, D., Zhang, M.: Broadband wavelength conversion based on parallel-coupled micro-ring resonators. IEEE Photon. Technol. Lett. **30**(17), 1559–1562 (2018)
15. Bharti, G.K., Biswas, U., Rakshit, J.K.: Design of micro-ring resonator based all optical universal reconfigurable logic circuit. Optoelectron. Adv. Mater.-Rapid Commun. **13**(7–8), 407–414 (2019)
16. Butt, M.A., Khonina, S.N., Kazanskiy, N.L.: A serially cascaded micro-ring resonator for simultaneous detection of multiple analytes. Laser Phys. **29**(4), 046208 (2019)

An Efficient Decision Fusion Scheme for Cooperative Spectrum Sensing for Cognitive Radio Networks

Prakash Chauhan, Sanjib K. Deka, and Nityananda Sarma

Abstract In this work, we propose an efficient decision fusion scheme for CSS, which achieves target cooperative probability of detection while maintaining a lower cooperative false alarm probability which eventually enhances the network throughput. The proposed scheme enables the fusion center (FC) to select secondary users (SUs) for decision fusion according to their reliability weights. For every SU, the reliability weight is computed by jointly considering the reporting information of SUs for past K time slots along with their detection and false alarm probabilities. The value of K is determined by computing the entropy on primary user (PU) activities observed on a designated channel over a certain period of time. Simulation-based study shows the efficacy of the proposed scheme in terms of reducing in cooperative probability of false alarm and improving the network throughput compared to conventional approach.

Keywords Cooperative spectrum sensing · Reliability · Efficient decision fusion · Cognitive radio networks

1 Introduction

Cooperative spectrum sensing (CSS) in cognitive radio networks (CRNs) has been proven as an effective technique to overcome the issues of individual spectrum sensing and achieves better detection accuracy. In CSS, geographically distributed SUs collaborate among themselves to make a group decision about sensing through utilizing the reported individual sensing information to a common node called as fusion

P. Chauhan (✉) · S. K. Deka · N. Sarma
Tezpur University, Napaam, Tezpur, Assam, India
e-mail: prakashc@tezu.ernet.in

S. K. Deka
e-mail: sdeka@tezu.ernet.in

N. Sarma
e-mail: nitya@tezu.ernet.in

J. K. Mandal et al. (eds.), *Applications of Internet of Things*, Lecture Notes in Networks and Systems 137, https://doi.org/10.1007/978-981-15-6198-6_7

center (FC). The main task of the FC is to select suitable SUs that would take part in fusion process and to perform fusion using appropriate fusion rule. In CSS, fusion can be performed by FC using soft fusion (or data fusion) or hard fusion (or decision fusion). Due to the low complexity and communication overhead, hard fusion is widely used in CSS [1, 2]. In a network, SUs located at different geographical locations suffer from different level of attenuation and fading effects. Therefore, every SU may not contribute equally in the process of decision fusion [3]. But, the key issue with classical hard fusion rule such as OR or AND is that they give equal priority to every SU during decision fusion irrespective of SUs' sensing quality.

In literature, several works have been presented [4–8], which addressed the issues of suitable SU selection for decision fusion to optimize the detection accuracy of CSS. The work discussed in [4] presents technique to select SUs based on their reliability for decision fusion where reliability of SUs were computed based on their detection results. However, in this work, the types of fusion rule employed by FC was not discussed. A distributed technique for CSS was presented in [5], which uses the OR rule for fusion, and SUs were selected for fusion depending upon their reliability while how SUs' reliability were computed is not elaborated. A distance-based reliable CSS algorithm was put forwarded in the work [6] where reliability of SUs were decided based on SUs' distance from PU. But, it is reported that there exists many factors such as hidden terminal problem, shadowing issues other than distance-based attenuation which affects the detection accuracy of SUs in spectrum sensing. A selective CSS approach was discussed in [7], whose main objective was to select an optimal set of SUs for cooperation to enhance detection accuracy. The work in [7] performs well for an environment where most of the network parameters such as PU transmission power, noise power, distances of SUs from PU, and path loss exponent are known beforehand.

From the above literature, it is observed that most of the existing works decide reliability of SUs based on their individual detection performance. But, in a real-time sensing, SUs suffer from different levels of noise, fading, and shadowing issues depending upon their geographical location and reporting of their individual sensing information and may become erroneous due to such environmental hazards. Thus, during the computation of SUs' reliability, consideration of accuracy of reported sensing results could play an important role toward achieving the target probability of detection. Furthermore, in a network like CRNs with interweave mode of communication, gathering knowledge of all network parameters are not always possible.

In this work, we propose an efficient decision fusion scheme for CSS for interweave mode CRN, which achieves target cooperative probability of detection while maintaining a lower cooperative false alarm probability. The proposed scheme adopts a censoring mechanism, which enables the FC to select SUs for fusion according to their reliability weights. For every SU, the reliability weight is computed by jointly considering the reported sensing information of SUs for past K time slots along with their detection and false alarm probabilities. The value of K is determined by computing the entropy on observed primary user (PU) activities on a designated channel over a certain period of time. Simulation-based study shows the efficacy of the proposed scheme in terms of reduction of cooperative probability of false alarm and improvement of throughput compared to conventional approach.

2 System Model and Assumptions

We consider a network comprising of N number of SUs, denoted by $\mathcal{N} = \{1, 2, \ldots, N\}$, a primary user (PU) channel, and a fusion center (FC), which has the responsibility to perform fusion activity. We assume that SUs are synchronized with PU, and PU accesses the channel in a discrete time-slotted fashion. Because of less complexity and ease of implementation, it is assumed that every SU performs local spectrum sensing using energy detection (ED) technique. ED can be described using two statistical hypotheses, namely H_0 and H_1, where H_0 and H_1 represent absence and presence of PU signal in the channel, respectively, and can be represented by (1).

$$Y(t) = \begin{cases} x(t), & H_0 \\ h(t).s(t) + x(t), & H_1. \end{cases} \tag{1}$$

where $Y(t)$, $h(t)$, $s(t)$, and $x(t)$ represent power of the received signal, channel gain, transmitted PU signal power, and zero-mean additive white Gaussian noise at tth time slot of a given channel, respectively. The decision of local spectrum sensing is taken by comparing the value $Y(t)$ with sensing decision threshold denoted by λ_s. For a SU i, the decision of local spectrum sensing at time slot t denoted by $d_{i,t}$ is given by (2).

$$d_{i,t} = \begin{cases} 0, & \text{if } Y(t) < \lambda_s. \\ 1, & \text{else} \end{cases} \tag{2}$$

The performance of spectrum sensing operation performed by SUs is measured in terms of two metrics, namely probability of detection (P_d) and probability of false alarm (P_f), where P_d indicates the detection of PU signal on a channel as present when it is actually present and P_f indicates about detecting PU signal on a channel as present when it is actually absent. For a SU i, P_f and P_d are computed according to [9].

After local sensing is performed, SUs report their sensing information to the FC by using a dedicated common control channel. Once FC receives all the sensing information from the SUs, FC starts fusing them to make a global decision about sensing. In our work, we consider OR fusion rule for decision fusion since it is well suited to minimize interference to PU communication. The cooperative probability of detection and probability of false alarm for a cooperative group, G, having L SUs are given by (3) and (4), respectively.

$$P_{d,G} = 1 - \prod_{i=1}^{L}(1 - P_{d,i}) \tag{3}$$

$$P_{f,G} = 1 - \prod_{i=1}^{L}(1 - P_{f,i}) \tag{4}$$

where $P_{d,i}$ and $P_{f,i}$ represent the probability of detection and probability of false alarm of SU i.

3 Proposed Scheme

The proposed scheme selects SUs by determining the reliability weight (β) for each of the SUs by jointly considering their quality of reported individual sensing decision on past K time slots ($K \neq 0$) and detection probabilities. Here, quality of reported sensing decision of SU i denoted by Q_i refers to the probability of SU's reported sensing decision matched with global decision by FC. The detection probability of SU i indicates their individual probability of detection and false alarm, i.e., $P_{d,i}$ and $P_{f,i}$. Thus, Q_i can be determined using (5).

$$Q_i = \frac{\sum_{j=1}^{K} z_{i,j}}{K} \tag{5}$$

where $z_{i,j}$ be a binary variable whose value is one only when the individual reported sensing decision of SU i get matched with global decision of the FC at time slot j, otherwise zero and can be represented by (6).

$$z_{i,j} = \begin{cases} 1, & \text{if } d_{i,j} = d_{FC,j} \\ 0, & \text{else.} \end{cases} \tag{6}$$

Here, $d_{i,j}$ and $d_{FC,j}$ represent individual sensing decision of SU i and global sensing decision by FC at time slot j. Here, in the process to determine the value of Q_i, the value of K plays an important role.

3.1 Determination of K

The value of K indicates how many previous time slot's reported sensing information of SUs should be stored in buffer by FC to determine Q. Thus, K specifies the size of the buffer (B) at FC, which contains the reported sensing decisions of the SUs of previous time slots. While estimating K, we use the intuition that the buffer B must

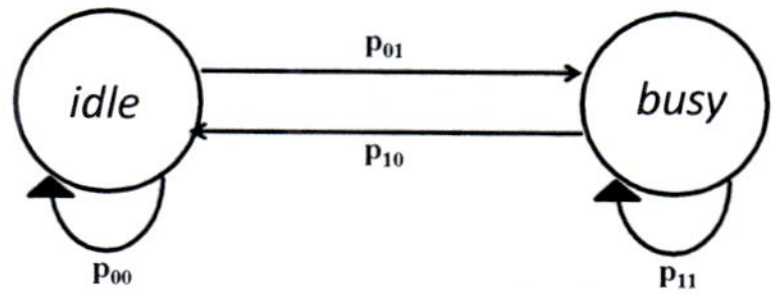

Fig. 1 DTMC based PU activity model

contain sufficient information, which will substantially capture the dynamics of PU activities on the channel. The dynamics of PU activities, i.e., PU presence or absence, on a given channel impacts the local sensing results of SUs. To determine K, we first observe the PU behavior on a given channel and compute the prior probability of PU activities. Inspired by [10], we consider that PU activities over a channel follow two-state discrete time Markov chain (DTMC) model, which is a framework for modeling PU activity in CRN due to the Markov property which gets exhibited during PU's channel usage pattern. For DTMC model, presence and absence of PU activity on a channel denoted by $busy$ (or 1) and $idle$ (or 0) state, respectively, are shown in Fig. 1.

Here, p_{00} and p_{01} refer to state transition probabilities from $idle$ to $idle$ and $idle$ to $busy$ state, respectively. Similarly, p_{10} and p_{11} represent state transition probabilities from $busy$ to $idle$ and $busy$ to $busy$ state, respectively. The state transition probabilities can be determined by performing long-term PU activity history observation on a given channel [11]. After computing the prior probabilities of PU activity in terms of state transition probabilities, we compute maximum entropy [12] on resultant probability to decide K which could capture the dynamics of PU activities accurately. The entropy of PU activities over w number of time slots denoted by ρ_w can be computed using (7).

$$\rho_w = -\sum_{u=0}^{1}\sum_{v=0}^{1} p_{uv}^{w} \log p_{uv}^{w} \tag{7}$$

where p_{uv}^{w} indicates state transition probability of PU from state u to state v which is computed when the PU activity is observed over a channel considering w number of time slots. Let us consider that a long-term past PU activity is observed over a given channel for M number of slots. Using these observations, K can be determined using (8).

$$K = f_w(\max(\rho_w)), \forall w \in \{o, o+1, o+2, \ldots, M\} \tag{8}$$

where $f_w()$ is a function which returns the value of w for which ρ_w is maximum and o being the initial value of w. Thus, equation (8) determines the value of K for which entropy is maximum so that the buffer B with size K can capture the dynamics of PU efficiently. FC uses the buffer B to store the K most recently reported sensing information of SUs in order to compute Q, which eventually helps in determining the reliability weight β.

3.2 Determination of β

In order to maintain a given target cooperative probability of detection with minimum cooperative probability of false alarm, SUs having with high Q values, high P_d, and low P_f can be considered as better candidates in decision fusion process. Thus, β is computed by (9).

$$\beta_i = \frac{Q_i P_{d,i}}{P_{f,i}}, \quad \forall i \in \mathcal{N} \tag{9}$$

3.3 Selection of Secondary Users (SUs) for Decision Fusion

To select the most appropriate SUs for decision fusion in CSS, we introduce a greedy-based approach, which selects SUs in descending order of their β until target cooperative probability of detection ($\overline{P_d}$) gets satisfied. Once $\overline{P_d}$ is entertained, the algorithm terminates and computes the cooperative probability of false alarm using (4). The average throughput C_i of SU i for a group G can be computed by (10) [2].

$$C_i = P_{H0}(1 - \frac{t_s}{T} - \frac{t_r}{T} - \frac{t_d}{T})(1 - P_{f,G})r_i, \quad \forall i \in \mathcal{N} \tag{10}$$

where P_{H0}, T, t_s, t_r, t_d, and r_i represent probability of channel being actually idle, length of a slot, sensing time, reporting time, decision time, and data rate of SU i, respectively. The steps of the greedy approach are given in Algorithm 1.

Algorithm 1: Greedy-based Secondary Users Selection

Input: $\mathcal{N}$, $\overline{P_d}$, β
Output: List of selected SUs

Step 1: Prepare the preference list of SUs, $\mathcal{L}$, in decreasing order of β

Step 2: SU selection round:
- 2.1: Form an empty cooperating group G, which contains the SUs those will be selected for decision fusion
- 2.2: Select SU l sequentially from top of the list $\mathcal{L}$ and insert in G
- 2.3: If no more SUs exists ($\mathcal{L}$ is exhausted), goto step 2.7, otherwise goto step 2.4
- 2.4: Compute the cooperative probability of detection for G, i.e., $P_{d,G}$
- 2.5: Check if target cooperative probability of detection is achieved or not (i.e., if $P_{d,G} >= \overline{P_d}$ or not). If yes, goto step 2.6. Otherwise, goto step 2.2
- 2.6: Stop SU selection and goto step 3
- 2.7: Target probability of detection can not be achieved. Exit.

Step 3: Compute $P_{f,G}$ and C.

Time complexity: The time complexity to prepare the SUs preference list is of order $O(N \log N)$, which is performed in step 1. The time complexity of step 2 for selection of N SUs for decision fusion is of order $O(N)$, and time complexity of step 3 is $O(N)$. Hence, the overall time complexity of the Algorithm 1 is $O(N + N + N \log N)$, which is dominated by $O(N \log N)$.

4 Performance Evaluation

The performance of the proposed scheme is evaluated in terms of minimization of cooperative probability of false alarm and hence enhancement of throughput using a MATLAB based simulation setup. The performance of the proposed scheme is compared with conventional scheme [1] in which fusion of SUs' decisions are done without considering reliability of SUs. For simulation, we consider $N = 50$, $\overline{P_d} = 0.9$, $P_{H0} = 0.5$, $T = 10$ s, $t_s = 1$ s, $t_r = 100$ ms, $t_d = 10$ ms, and SNR within a range of -40 to -24 db.

Figure 2a shows result of experiment conducted to determine K through computing the average entropy of PU activities versus w. As shown in the figure, the value of average entropy changes for different values of w, and it attains maximum value while $w = 400$. Thus, the value of w for which average entropy becomes maximum is selected as K to determine the value of β for each of the SUs. Figure 2b shows the relationship between SNR of PU signal vs cooperative probability of false alarm. The figure indicates that with the decrease in SNR level, cooperative probability of false alarm for a group G increases for the conventional as well as for the proposed scheme. However, the proposed scheme outperforms the conventional scheme and shows approximately 3.5 times lower false alarm probability when SNR = -40 db. This is because in the proposed scheme SUs are selected according to their β values, and therefore, SUs with high sensing accuracy got selected for decision fusion.

Figure 3a reveals that with the increase in target detection probability, the cooperative probability of false alarm also increases for both the schemes. When target

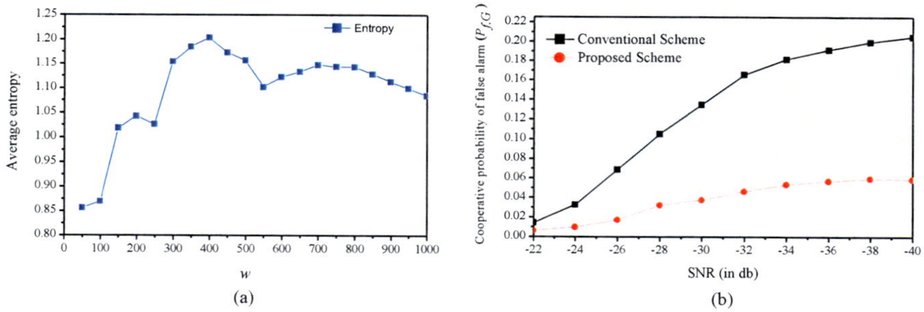

Fig. 2 **a** Average entropy versus w and **b** SNR versus cooperative probability of false alarm, $\overline{P_d} = 0.9$

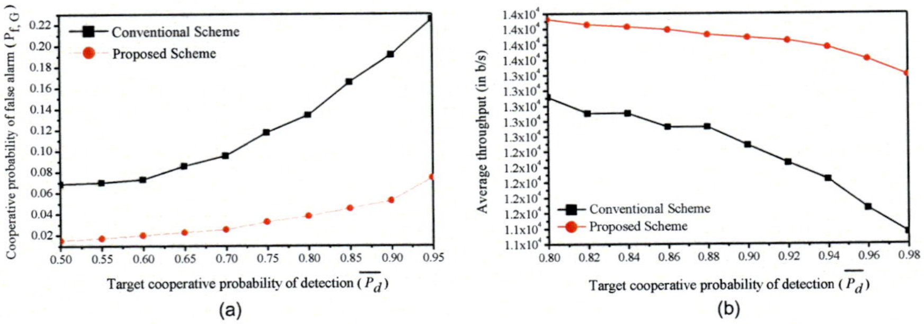

Fig. 3 Target cooperative probability of detection versus **a** cooperative probability of false alarm and **b** average throughput, when SNR = −30 db

probability of detection increases, more number of SUs need to take part in decision fusion process and as a result the cooperative false alarm probability rises higher. As the proposed scheme selects SUs according to their β value, it shows an average 3.57 times lower false alarm probability compared to the conventional scheme when target detection probability varies from 0.5 to 0.95.

Finally, Fig. 3b reveals that with the increase in target detection probability, the achieved average throughput by SUs decreases for both the schemes. This happens because as the target detection probability increases, the cooperative probability of false alarm increases; since the throughput is the complimentary function of probability of false alarm, higher false alarm impacts the probability of channel utilization to be lower by SUs, which eventually leads to lower throughput. Further, the proposed scheme outperforms the conventional scheme in terms of average throughput achieved by SUs through cooperation.

5 Conclusion

In this work, we proposed an efficient decision fusion scheme for cooperative spectrum sensing for cognitive radio networks using which target cooperative probability of detection was achieved maintaining a reasonably low cooperative false alarm probability, which eventually enhanced the average throughput of the network. In the proposed scheme, a censoring-based mechanism was presented for SU selection during decision fusion. Simulation results showed that the proposed scheme outperforms the conventional scheme of decision fusion both in terms of reducing the cooperative probability of false alarm and reasonably enhancing the average throughput achieved by SUs through cooperation.

References

1. Saad, W., Han, Z., Debbah, M., Hjorungnes, A., Basar, T.: Coalitional games for distributed collaborative spectrum sensing in cognitive radio networks. In: INFOCOM 2009, pp. 114–2122. IEEE, New York (2009). https://doi.org/10.1109/INFCOM.2009.5062135
2. Deka, S.K., Chauhan, P., Sarma, N.: Constraint based cooperative spectrum sensing for cognitive radio network. In: International Conference on Information Technology, 2014, pp. 63–68. IEEE, New York (2014). https://doi.org/10.1109/ICIT.2014.12
3. Cui, T., Kwak, K.S.: Cooperative spectrum sensing with adaptive node selection for cognitive radio networks. Wireless Personal Commun. **74**(4), 1879–1890 (2014). https://doi.org/10.1007/s11277-014-2050-2
4. Wang, L., Zhang, S., Gu, L.: Reliability-based cooperative spectrum sensing algorithm in cognitive radio networks. In: 24th International Conference on Software, Telecommunications and Computer Networks (SoftCOM), pp. 1–5. IEEE, New York (2016).https://doi.org/10.1109/SOFTCOM.2016.7772176
5. Gupta, J., Chauhan, P., Nath, M., Manvithasree, M., Deka, S.K., Sarma, N.: Coalitional game theory based cooperative spectrum sensing in CRNS. In: 18th International Conference on Distributed Computing and Networking (ICDCN). ACM, New York (2017).https://doi.org/10.1145/3007748.3007759
6. Verma, G., Sahu, O.P.: A distance based reliable cooperative spectrum sensing algorithm in cognitive radio. Wireless Personal Commun. **99**(1), 203–212 (2018). https://doi.org/10.1007/s11277-017-5052-z
7. Dhurandher, S.K., Woungang, I., Gupta, N., Jain, R., Singhal, D., Agarwal, J., Obaidat, M.S.: Optimal secondary users selection for cooperative spectrum sensing in cognitive radio networks. In: IEEE Globecom Workshops (GC Wkshps), pp. 1–6. IEEE, New York (2018).https://doi.org/10.1109/GLOCOMW.2018.8644208
8. Hussain, A.S., Deka, S.K., Chauhan, P., Karmakar, A.: Throughput optimization for interference aware underlay CRN. Wireless Personal Commun. pp. 1–16 (2019). https://doi.org/10.1007/s11277-019-06257-6
9. Hao, X., Cheung, M.H., Wong, V.W.S., Leung, V.C.M.: A coalition formation game for energy-efficient cooperative spectrum sensing in cognitive radio networks with multiple channels. In: Global Telecommunications Conference (GLOBECOM 2011), pp. 1–6. IEEE, New York (2011). https://doi.org/10.1109/GLOCOM.2011.6134135
10. Gelabert, X., Sallent, O., Pérez-Romero, J., Agustí, R.: Spectrum sharing in cognitive radio networks with imperfect sensing: a discrete-time Markov model. Comput. Networks **54**(14), 2519–2536 (2010). https://doi.org/10.1016/j.comnet.2010.04.005
11. Csurgai-Horváth, L., Bitó, J.,: Primary and secondary user activity models for cognitive wireless network. In: 11th International Conference on Telecommunications, pp. 301–306. IEEE, New York (2011)
12. Jaynes, E.T.: Probability Theory: The Logic of Science. Morgan Kaufmann, San Francisco (2003)

Detection of Early Breast Cancer Using A-Priori Rule Mining and Machine Learning Approaches

Anwesha Banik, Birajit Debbarma, Monalisha Debnath, Sun Jamatia, and Ankur Biswas

Abstract In today's world, breast cancer is extremely predominant in females that establishes in the breast and further extends to other locales of the body in the track of time. It is the second major ailment that causes decease. In long term, an early detection can reduce the death rate due to breast cancer appreciably. The crucial point for early prediction is to recognize the cancer cells at virgin stages. Various researches are carried out on breast cancer detection using mammography, ultrasounds, CT scans, PET, MRI, biopsy, etc. Still, these techniques are expensive, prolonged and sometimes unsuitable for young females. Hence, a fast and accurate detection system is highly demanded. In recent years, data mining and machine learning techniques are given utmost attention for early stage breast cancer detection. The aim of this paper is to present a framework for accurate and quick conclusion of breast cancer using machine learning techniques. We applied our proposed technique on SEER dataset of breast cancer and obtained highly appreciable results with accuracy of 99.9% using random forest. Various rules are also presented in support of breast cancer detection using A-priori algorithm.

Keywords Breast cancer · Machine learning · Data mining · Random forest · A-priori

A. Banik · B. Debbarma · M. Debnath · S. Jamatia · A. Biswas (✉)
Department of Computer Science and Engineering, Tripura Institute of Technology, Narsingarh 799009, Tripura, India
e-mail: abiswas.tit@gmail.com

A. Banik
e-mail: banikanwesha00@gmail.com

B. Debbarma
e-mail: birajitdebbarma@gmail.com

M. Debnath
e-mail: monalishadebnath7@gmail.com

S. Jamatia
e-mail: sunjamatia@gmail.com

J. K. Mandal et al. (eds.), *Applications of Internet of Things*, Lecture Notes in Networks and Systems 137, https://doi.org/10.1007/978-981-15-6198-6_8

1 Introduction

As on 2019, cancer is the foremost global public health issue and the second biggest cause of death in the USA [1]. Breast cancer is one form of it that occurs when the cells in the breast grow in an uncontrolled manner. The breast cancer is the mostly common occurring cancer in women. In 2019, 268,600 new cases of female breast cancer are expected to occur and an additional 41,760 women will die from this disease [2]. It is the second most common disease in most towns and second most common in rural areas of India [3] as per National Cancer Registry Program (http://www.breastcancerindia.net/statistics). This accounts for 25–32% of all female cancers, which means 1/4 (or even probable 1/3) of all female cancers. Typically, the cancer forms in either the lobules or the ducts of the breast. A breast cancer may be invasive or non-invasive. Breast cancer takes place as cancer cells inside milk ducts or lobules break into surrounding breast tissue. There are many forms of invasive breast cancer, but invasive ductal carcinoma and lobular carcinoma are the most common ones. In the milk ducts or lobules in the breast, non-invasive cancers remain. The tissues in the breast or elsewhere do not expand into or invade normal tissues. Non-invasive carcinoma is often referred to as in situ (in the same place) or pre-cancer. An early classification of type of cancer that the patient having is the crucial step in diagnosis because it will provide early treatment to the patient, and the cancer may remain confined in those places where they are detected.

A prediction system is needed that will assist doctor to predict the type of cancer more efficiently. Hence, in this paper, a diagnosis and prediction system using data mining and machine learning techniques is proposed. Data mining extracts or learns the pattern of occurring particular disease from a large data source and apply this pattern to predict the outcome of new patient. Various information and rules are revealed in support of breast cancer detection using data mining 'A-priori' algorithm is presented in this paper. For early prediction, random forest classifier is chosen, 15,774 instances from surveillance, epidemiology, and end results (SEER) dataset [4] is used to train the model. The proposed system will facilitate practitioners for conclusion of breast cancers with minimal tests in short time. It will also help in finding various statistics and discovering hidden patterns for breast cancer detection. Various patterns are revealed related to breast cancer and an approach is taken to find out best classifier for prediction. The remaining paper is as follows: In Sect. 2, we present background literature of breast cancer detection. Section 3 presents methodology of the proposed system, whereas Sect. 4 demonstrates the results obtained from the proposed system. And finally, Sect. 5 states the concluding remarks and some future directions in this research fields.

2 Literature Survey

Several literatures are available on breast cancer detection using SVM through selection and classification, genetic algorithms, ANN achieving higher accuracy [5, 6]. Modified ANN was also utilized for classifying breast cancer, splitting the dataset into two classes: benign and malignant [7]. Researches on SEER database for prediction of breast cancer survivability using decision tree algorithms obtained accuracy of 0.7678 [8]. The ANN and C5 decision tree to expand the prediction model was set down by Delen et al. [9]. C5 offered 93.6% precision while ANN provided 91.2% accuracy. In this section, different techniques and means used in detection of breast cancer are discussed.

2.1 Machine Learning

It is a subset originated from artificial intelligence can be an alternative option for researches on breast cancer. Through training also called as 'learning' huge dataset, machine learning intends to provide stout models being capable of predicting results of other unknown datasets. In medical domain, particularly, the study of cancer, the contribution of rapidly built up genomic data and databases from clinics are remarkable in a variety of applications of machine learning [10]. It has supported the prediction of cancer vulnerability, reappearance and survival by learning through mammography, genomic and clinical features [11]. The SEER dataset assembles information on incidence and survival covering a significant portion of US population has proven to be an important means to predict survival of numerous cancers like breast and lung cancer [12–15].

2.2 Random Forest

A random forest is a classification technique that consists of multiple random regression trees. The output of the multiple trees is combined to generate the aggregated results of regression. A random variable is also utilized to decide the split point when every single tree is created, for example, the location of the dividing coordinate and the separating point. A different subgroup of random variables is considered by every other tree. The prediction in random forest is shown in Fig. 1.

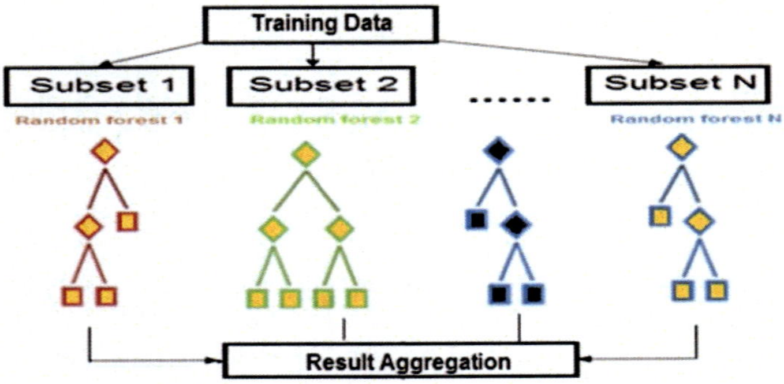

Fig. 1 Random forest classification

2.3 A-Priori Algorithm

Association rule is a technique in data mining that guarantees that data analysis finds association pattern. The trends observed indicate the relation among the attributes of dataset values that appear frequently. A-priori has a redundant structure whose objective is to discover the strongest rule in the dataset. The dataset is searched multiple times for this purpose. The number of repeats in the data is determined in the first step which represents the support factor. Data below that factor are excluded in itemset. Candidate itemsets are created in each iteration. Iterations are continued until no itemsets are found.

3 Material and Method

This paper is concerned about discovering effective technique for predicting the breast cancer through comparing various predictive models and to find the best one based on previous patient clinical records. Following machine learning algorithms are applied in this paper: (1) J48, (2) Naive Bayes (3) random forest. To evaluate the performance of these models, SEER dataset of Program of the National Cancer Institute (NCI) is used. The SEER program collects and releases de-identified data for individual cancer diagnoses and outcomes in the USA. SEER gathers cancer case reports from various sites and sources around the USA. The compilation of data began in 1973. This dataset consists of cases from 1975 to 2015. It consists of nearly 8 lakhs patient data and 72 attributes which are the main reason of using this dataset. After preprocessing, the number reduced to 3 lakhs and 13 attributes. To estimate the test error of each model, tenfold cross-validation method is implemented. The overall flow diagram is shown in Fig. 2.

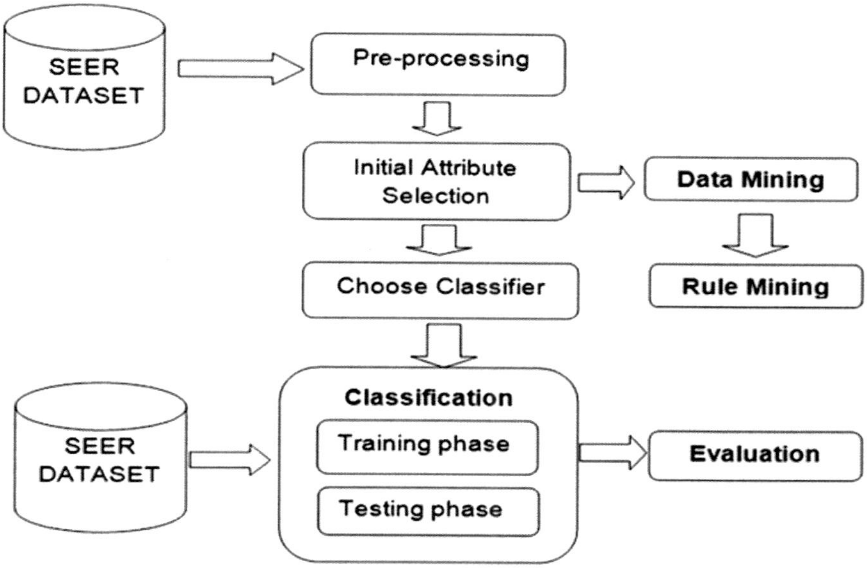

Fig. 2 Flow diagram of the methodology

3.1 *Preprocessing*

Data preprocessing is important step in data mining. For further review, raw input data should be translated into the correct format. The preprocessing task comprise data fusing from several source, data cleaning to eliminate noisy and redundant observation, and selection of record and characteristic that are appropriate for data mining. Preprocessing is being applied on 'SEER' dataset to make it compatible for data mining

Initially, the attributes or columns having all values missing are removed. From this, attribute count dropped from 72 to 70. Secondly, the attributes like 'patient-id', 'registry_id', 'marital_status', 'Race', which are not related breast cancers are recognized and removed. Thirdly, redundant attributes are removed. For example: 'behaviour1' and 'histology1' are redundant attributes of 'behaviour2' and 'histology2'. ICCC site recode ICD-O-3/WHO 2008, ICCC site recode extended, ICD-O-3/WHO 2008, Behavior Recode for Analysis, etc., are all redundant attributes. EOD_TUMOUR_SIZE and CS_Tumor_Size have information on tumor size from 1975 to 2003 and 2004 to 2015 diagnosis years, respectively. CS_Tumour size is kept as it consists of information on recent years. All the instances where 'CS_TUMOUR size' is missing are removed. EOD_TUMOUR_SIZE is also removed. Since, this paper is concerned about predicting the type of cancer, than cancer staging system attributes are irrelevant. Hence, this attributes ('Derived AJCC T', 'Derived AJCC N', 'Derived AJCC M', 'Derived AJCC Stage Group') are dropped. After this we are left with 28 columns. Finally, columns having

80% null values are removed. Finally, we are left with 13 attributes, viz. (Sex, age_at_diagnosis, Sequence_Number, 'Year_of_diagnosis', 'Primary_Site', Laterality, Histology2, Behavior, 'Grade', Diagnostic_Confirmation, 'CS_Extension', Regional_Nodes_Positive, 'CS_Tumor_Size')

3.2 Data Mining Through Weka

The data must be processed in a manner so that it is appropriate for future analysis, hence, the basic set of data converted in .csv format or .arff format suitable for data mining and classification. After data gets loaded in Weka, it illustrates information of pre-selected attributes such as total attribute number and sum of weight. The majority of the attributes are numeric or alphanumeric. As per data mining classification compatibility in Weka, the desired attribute is transformed to nominal values. From our analysis, we can conclude that breast cancer increases in women after 45 years of age. Analysis is performed on 'age_at_diagnosis' attribute is shown in Fig. 3.

A. Analysis is done on primary site attribute which provides information from where the tumor is originated. A breast is divided into four quadrants (upper inner quadrant, upper outer quadrant, lower inner quadrant and lower outer quadrant). These quadrants are coded as per standard.
C500: nipple area, C501: central portion of breast (subareolar) area extending 1 cm around areolar complex, C502: upper inner quadrant (UIQ) of breast, C503: lower inner quadrant (LIQ) of breast, C504: upper outer quadrant (UOQ) of breast, C505: lower outer quadrant (LOQ) of breast, C506: auxiliary tail of breast, C508: overlapping lesion of breast, C509: breast, NOS.
It has been founded that in 'upper_quadrant' breast cancer occur first as shown in Fig. 4.
B. Analysis is carried out for finding the number of tumors occurring on upper quadrant is invasive. From analysis, we can conclude that 80% cases occurring

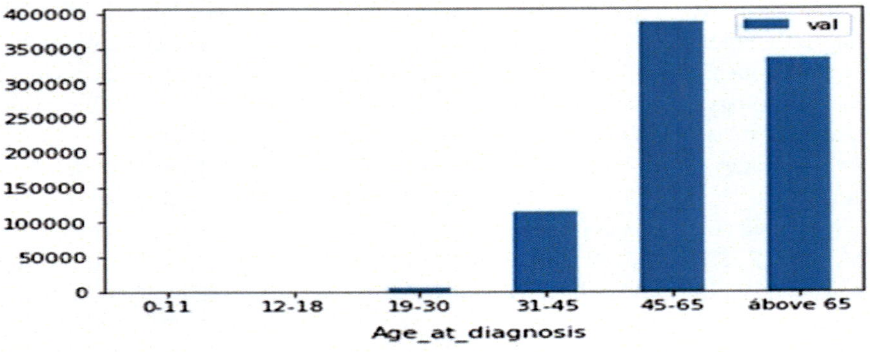

Fig. 3 Analysis of 'age_at_diagnosis'

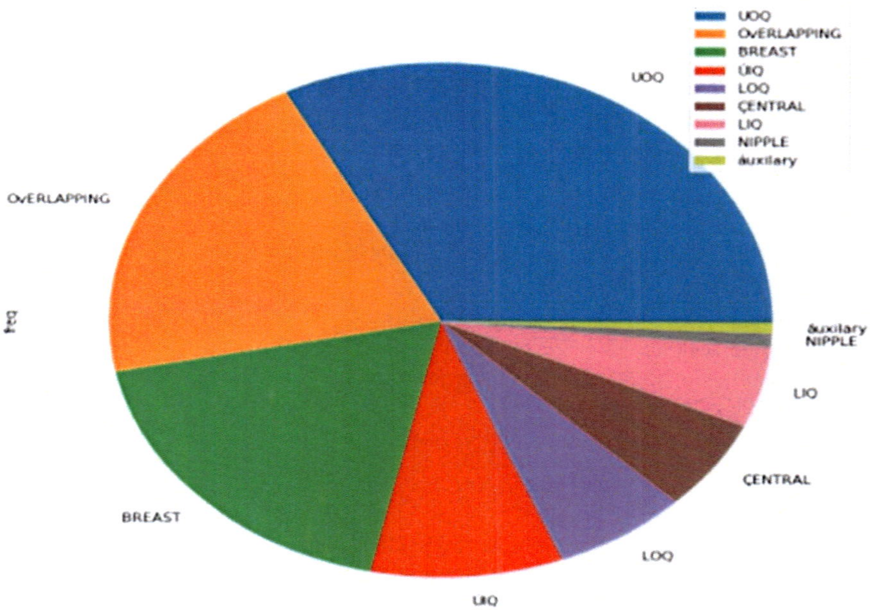

Fig. 4 Analysis of 'Primary Site' attributes

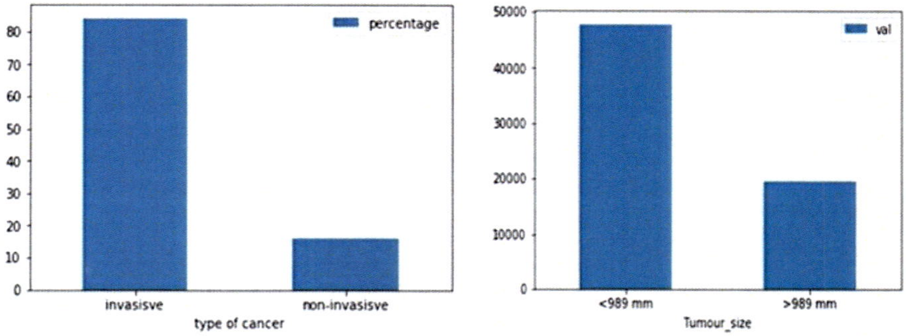

Fig. 5 Analysis for tumor size and invasive

on upper quadrant is invasive and tumor size less than 989 mm is non-invasive in nature as shown in Fig. 5.

3.3 *Classification*

Data available in .arff is applied for random forest (RF) classification. The RF classification requires the training set to be arranged to train the model competent to group the data instance into recognized class. The classification procedure comprises of the subsequent steps: building the training dataset, classification of class attributes, classification of appropriate attributes, model learning from training data and finally test data classification using learned model.

A. **Training Phase**: In this phase, random 1000 records are selected from 1597 instances of pre-processed data to achieve the classification rule set using random forest. To verify, other classification algorithms like j48 and Naïve Bayes are further applied. The results are compared and recoded that random forest performs best.
B. **Testing Phase**: In this phase, all the three classifier has been applied on the whole dataset of 1183 records. The outcome of actual and predicted values attained through classification is represented in the confusion matrix.

4 Results

This section demonstrates the prediction system using diverse classification algorithms, like J48, Naive Bayes and RF classifier. The classification is an associated machine learning procedure to forecast the membership for groups of instances. RF classification performed best among the other algorithms available and achieved classification accuracy of (99.96%) on 15,774 training instances. The accuracy obtained by J48 and Naive Bayes on same set of instances is (99.27%) and (98.52), respectively. The overall summaries of three classifiers are presented in Table 1.

The detailed accuracy of random forest classifier is shown in Table 2 with confusion matrix in Table 3 exhibits the correctness of J48, Naive Bayes and random forest, respectively.

Table 1 Classification outline of different models

Parameters	J48	Naive Bayes	Random forest
Correct classification	15695 (99.27%)	15542 (98.52)	15769 (99.96%)
Incorrect classification	115 (0.729%)	232	05 (0.03%)
Kappa statistic	0.5003	0.4194	0.9852
Mean absol_error	0.0041	0.0055	0.0016
RMSE	0.0455	0.0579	0.0175
Relative absol_error	65.49%	86.323%	24.85%
Root relative squared error	81.61%	103.77%	31.42%
Instances	15774	15774	15774

Table 2 Detailed accuracy by random forest

	TP rate	FP Rate	MCC	ROC area	PRC area	Class
	1.000	0.029	0.985	1.000	1.000	1
	0.933	0.000	0.966	1.000	0.998	2
	1.000	0.000	1.000	1.000	1.000	4
	1.000	0.000	1.000	1.000	1.000	6
	1.000	0.000	1.000	1.000	1.000	7
	1.000	0.000	1.000	1.000	1.000	8
	1.000	0.000	1.000	1.000	1.000	9
Weighted avg.	1.000	0.029	0.985	1.000	1.000	

Table 3 Confusion matrix

a	b	c	d	e	f	g	← classified as
15601	0	0	0	0	0	0	a = 1
5	70	0	0	0	0	0	b = 2
0	0	3	0	0	0	0	c = 4
0	0	0	3	0	0	0	d = 6
0	0	0	0	23	0	0	e = 7
0	0	0	0	0	4	0	f = 8
0	0	0	0	0	0	65	g = 9

Moreover, the prediction system also derived some rules through rule mining using A-priori algorithm. The following statistics are the strongest rules for breast cancer:

A-Priori Algorithm

Min. support = 0.9 (14,197 instances), Min_metrics <confidence>: 0.9
Cycle numbers = 2
Large itemset generated = Itemsets: L(1) = 3, L(2) = 3, L(3) = 1

The obtained best rules are represented as,

1. CS Mets at Dx = 0 14747 ⇒ SEER Type of Follow-up = 214747 <conf:(1)> lift:(1) lev:(0) [56] conv:(56.09)
2. Diagnostic Confirmation = 1 CS Mets at Dx = 0 14691 ⇒ SEER Type of Follow-up = 2 14691 <conf:(1)> lift:(1) lev:(0) [55] conv:(55.88)
3. Diagnostic Confirmation = 1 15601 ⇒ SEER Type of Follow-up = 2 15599 <conf:(1)> lift:(1) lev:(0) [57] conv:(19.78)
4. CS Mets at Dx = 0 14747 ⇒ Diagnostic Confirmation = 114691 < conf:(1) > lift:(1.01) lev:(0.01) [105] conv:(2.84)

5. CS Mets at Dx = 0 SEER Type of Follow-up = 2 14747 ⇒ Diagnostic Confirmation = 114691 <conf:(1)> lift:(1.01) lev:(0.01) [105] conv:(2.84)
6. CS Mets at Dx = 0 14747 ⇒ Diagnostic Confirmation = 1 SEER Type of Follow-up = 2 14691 <conf:(1)> lift:(1.01) lev:(0.01) [107] conv:(2.87)
7. SEER Type of Follow-up = 2 15714 ⇒ Diagnostic Confirmation = 1 15599 <conf:(0.99)> lift:(1) lev:(0) [57] conv:(1.49)
8. Diagnostic Confirmation = 1 SEER Type of Follow-up = 2 15599 ⇒ CS Mets at Dx = 0 14691 <conf:(0.94)> lift:(1.01) lev:(0.01) [107] conv:(1.12)
9. Diagnostic Confirmation = 1 15601 ⇒ CS Mets at Dx = 0 14691 <conf:(0.94)> lift:(1.01) lev:(0.01) [105] conv:(1.11)
10. Diagnostic Confirmation = 1 15601 ⇒ CS Mets at Dx = 0 SEER Type of Follow-up = 2 14691 <conf:(0.94)> lift:(1.01) lev:(0.01) [105] conv:(1.11).

5 Conclusion and Future Scope

An early exposure of breast cancer can decrease the mortality appreciably in the long term. The goal in this paper was to build a predictive model for breast cancer detection using data extraction and machine learning methods through relevant attributes of SEER dataset. An attempt was made to classify SEER dataset of breast cancer and provide new patterns for attributes 'age_at_diagnosis', 'Primary_Site', 'CS_Tumor_size' using random forest algorithm. And it is also tried to predict the existence of breast cancer with training set of samples of 15,774 records and further apply the acquired rules of classification on the entire dataset. An accuracy of 99.9% for training data is appreciable. The performance of algorithm is also compared with other classification techniques. It is also proof of the effective use of machine learning or data mining techniques to predict breast cancer. The conclusions of this paper should be used by oncologists as a method to establish accurate breast cancer diagnostics.

Future enhancement of this work includes improvisation of the random forest algorithm to improve the classification rate to achieve greater accuracy. SEER data with 13 attributes has been included in all the validations made in this study. A more study with other attributes of specific parameter settings would be performed to strengthen the prediction model as well as to establish new capacities. Furthermore, random forest implementations should be extensively tested. Inconsistency of data, the presence of missing values, noisy data and outliers are the big problem in data mining and machine learning. Statistical and machine learning approaches must also be used for data quality control.

References

1. Siegel, R.L., Miller, K.D. Jemal, A.: CA Cancer J. Clin. **69**(1), 7–34 (2019)
2. Breast Cancer: Statistics, American Society of Clinical Oncology (ASCO), (2019)

3. Indian Council of Medical Research, Department of Health Research: Ministry of Health & Family Welfare, Government of India, Media Report (2019)
4. SEER Dataset: Surveillance, Epidemiology, and End Results (SEER) Program (www.seer.cancer.gov) Research Data (1973–2008), National Cancer Institute, DCCPS, Surveillance Research Program, Cancer Statistics Branch, released April 2011, based on the November 2010 submission. www.seer.cancer.gov
5. Purnami, S.W., Rahayu, S.P., Embong, A.: Feature selection and classification of breast cancer diagnosis based on support vector machine. IEEE (2008)
6. Lambrou, A., Papadopoulos, H., Gammerma, A.: Evolutionary conformal prediction for breast cancer diagnosis. In: Proceedings of the 9th International Conference on Information Technology and Applications in Biomedicine (2009)
7. Keivanfard, F., Teshnehlab,M., Shoorehdeli, M.A.: Feature selection and classification of breast cancer on dynamic magnetic resonance imaging by using artificial neural networks. In: Proceedings of the 17th Iranian Conference of Biomedical Engineering (ICBME2010) (2010)
8. Ya-Qin, L., Cheng, W., Lu, Z.: Decision tree based predictive models for breast cancer survivability on imbalanced data. IEEE (2009)
9. Delen, D., Walker, G., Kadam, A.: Predicting breast cancer survivability: comparison of three data mining methods. Artif. Intell. Med. **34**, 113–127 (2005)
10. Obermeyer, Z., Emanuel, E.J.: Predicting the future—big data, machine learning, and clinical medicine. N. Engl. J. Med. **375**(13), 1216–1219 (2016)
11. Kourou, K., Exarchos, T.P., Exarchos, K.P., Karamouzis, M.V., Fotiadis, D.L.: Machine learning applications in cancer prognosis and prediction. Comput. Struct. Biotechnol. J. **13**, 8–17 (2015)
12. Kim, J., Shin, H.: Breast cancer survivability prediction using labeled, unlabeled, and pseudo-labeled patient data. J. Am. Med. Inform. Assoc. **20**(4), 613–618 (2013)
13. Lynch, C.M., Abdollahi, B., Fuqua, J.D.: Prediction of lung cancer patient survival via supervised machine learning classification techniques. Int. J. Med. Inform. **108**, 1–8 (2017)
14. Lynch, C.M., van Berkel, V.H., Frieboes, H.B.: Application of unsupervised analysis techniques to lung cancer patient data. PLoS ONE **12**(9), e0184370 (2017)
15. Ayer, T., Alagoz, O., Chhatwal, J., Shavlik, J.W., Kahn, C.E., Burnside, E.S.: Breast cancer risk estimation with artificial neural networks revisited: discrimination and calibration. Cancer **116**(14), 3310–3321 (2010)

Effect of Linear Features to Determination of Sleep Stages Classification from Dual Channel of EEG Signal Using Machine Learning Techniques

Santosh Kumar Satapathy and D. Loganathan

Abstract Sleep disorder is nowadays affected by all generations of age groups. For proper diagnosis of sleep disorder, the basic important step is to analysis of sleep quality. Since the traditional manual sleep staging is time-consuming and due to more human interpretation its accuracy toward sleep stage classification is not accurate. Thus, currently researchers have used the automated process on sleep monitoring which ultimately support for sleep experts for analyzing abnormality occurred during sleep. The important objective of this research work is analysis the effect of linear features of PSG signals and how far their effectiveness for the best classification among the different stages of sleep states. EEG is suitable for sleep study because the EEG signal directly extracted from the brain, which is ultimately helpful for proper tracking on brain behavior. Here, we have considered two channels such as F3-A2 and C3-A2 of EEG signal and considered gender specified subjects. To characterize the sleep behavior of the subjects, we have obtained linear properties from the input channel. Here, we have focused on four basic work scenarios accuracy in terms of sleep stage classification: (i) Channel effectiveness, (ii) subject effectiveness, (iii) combination of feature selection effectiveness and (iv) classification effectiveness to discriminating the different sleep stages accurately. For scenario 1, it has reached the overall accuracy of 95.9% for the C3-A2 channel. According to scenario 2, female subject sleep stage classification has reached the overall accuracy of 95.9% for the C3-A2 channel through SVM classifier and for KNN it has to be reached 95.2% and for DT, it has achieved 94.8% overall correctness for identifying the sleep stages. Scenario 3, it has observed that for channel C3-A2 of the subject-18 male category the selected features for classifiers. Finally, in scenario 4, it has shown that the SVM classifier achieved the highest accuracy level to be discriminating against the transition of different sleep stages accurately. This study shows that how far which channel, subject, features and classifiers effectiveness toward the diagnosis

S. K. Satapathy (✉) · D. Loganathan
Pondicherry Engineering College, Puducherry, India
e-mail: santosh.satapathy@pec.edu

D. Loganathan
e-mail: drloganathan@pec.edu

J. K. Mandal et al. (eds.), *Applications of Internet of Things*, Lecture Notes in Networks and Systems 137, https://doi.org/10.1007/978-981-15-6198-6_9

of sleep diseases so that it make more suitable for scientific and clinical sleep disorder assessment and diagnosis.

Keywords Sleep stage classification · Electroencephalogram · Linear features · Machine learning classifier

1 Introduction

The first step toward analysis of irregularities on sleep patterns is analysis of sleep quality. For maintaining good human health, proper sleep to be maintained in night is a basic requirement and it directly control upon our mental health and physical health. In last 20 years, the major changes reflected in human life styles in day to day life and it has seen that the ratio of sleep-related diseases are sharply increased and their associated impacts in all age groups across the world. It has found from the survey report 2013–2014, conducted in the USA by National Health Agency, finally it has observed from survey that children under age 18 with accompanying single parents spend shorter sleep during night and this ratio is high incomparable to adults with two-parent and adults who living without any children [1]. The survey conducted in the year 2014 by National Sleep Foundation with subject to sleep-related diseases and it has found that 45% of Americans affected by low quality sleep and its associated diseases [2]. The Sleep Health Foundation (SHF) 2016 survey in Australia has found that the average sleep time is 7 h but according to survey report, we have observed that 76% who sleep less than 5½ h and also reported that maximum they have day time impairment and other sleep-related symptoms [3]. Sleep is one of the resting states for humans. In this state generally, humans are unconscious toward major activities happening in their surrounding environment. Investigation has been conducted to understand the different sleep processes for various purposes. From that study, it has been found that one purpose is the identification of sleep disorder and its associated major related diseases. Some sleep disorders cause threats like in the later part of our life such as obstructive sleep apnea, insomnia and narcolepsy [4]. The primary steps for diagnosis of any type of sleep-related diseases are analysis of sleep stages and its irregularities patterns [5]. The standard procedure to identify the sleep disorder is analysis of the sleep cycle and its sleep quality and for that one of the standard techniques is sleep scoring, this score to be extracted from subjects during sleep from fixed electrodes associated with the brain. This total procedure can be called as sleep test or polysomnography (PSG) test. In general, PSG test make an important role toward diagnosis of any type of sleep-related issues by considering three physiological signals such as electroencephalogram (EEG), electrooculogram (EOG) and electromyogram (EMG). Apart from these three signals sometimes the sleep experts have suggested to consider more other information such heart rhythm, respiratory airflow, blood oxygen saturation and other measurements. In this proposed study, we have considered these extracted recordings and are extracted from subject's different parts of the body from fixed electrodes. These recorded signals segmented into epochs and as per our proposed study, we have also taken 30s non-overlapping epochs for measuring the different sleep stages. All these recording procedures have to be monitored through a set of sleep experts and technicians. Since from 1957 to 2007, the sleep staging analysis done through Rechtschaffen & Kales (R&K) sleep rules, and after 2007, the sleep annotations done through AASM rules, which was slight modifications from R&K rules. Early 1957, the technicians have observed practically two phases of sleep that have to be identified such as the non-rapid eye movement (NREM) and the other stage is rapid eye movement (REM) [6]. In 1968, R&K has introduced a new sleep manually, according to which the non-rapid eye movements consisted of four sub-sleep stages such as NREM1, NREM2, NREM3, NREM4. Since 2007, the American Academy of Sleep Medicine redefined the sleep stages and re-declared new rules in sleep stage classification. As per AASM manuals, the NREM phase of sleep is divided into three stages such as NREM1, NREM2, NREM3. According to AASM manuals, NREM1 and NREM2 are the light sleep stages and N3 and R are the deep stages. Nowadays, all types of sleep problem research work are followed as per AASM manuals. For the proper diagnosis of sleep stages, the most appropriate signal for sleep is to EEG from PSG [7, 8]. In this study, we have considered the EEG signals and its extracted features for sleep scoring and also we have dealt with 30s segment epochs for this research work.

Generally, sleep experts explain different characteristics of sleep stages through analyzing the recorded electroencephalogram signals from subjects. The NREM1 stage is beginning of actually sleep. In this transition stage, the heart rates are normal with breathing parameters. Here, a person can be easily disturbed due to high sound, high temperature, etc. Generally, in this sleep step, the theta waves are found in the recorded signals. The sleep state starts from NREM2. Here, the overall muscle activities substantially decrease. In this step, a person is less conscious of outer activities and during this stage the major parts of sleep patterns are in nature of spindles and k-complexes. NREM3 stage termed as deep sleep stage. Here, a person has less awareness of the surrounding incidents. The REM sleep stage happened around 15–20% of whole sleep and this stage generally occurred before complete wakefulness. Generally, in this stage, the person has made more movements of eyes. Here, mixed frequency waves with low amplitude behavior are observed from recorded EEG signal. The PSG test considers the bio-signals from placed electrodes attached in the brain of patients. The electrodes include a combination of EEG, ECG, EMG and EOG [9]. Besides here, we have also mentioned the possible classification cases of sleep stages in Table 1, but in present research work, our focus toward two-state sleep stage classification to detect the sleep disorder.

The traditional sleep staging process completely depends on sleep expert's visual interpretations of extracted signals. In this visual interpretation approach, certain disadvantages occurred due to huge bulks of data to monitor and it takes more time to visualize the recorded wave patterns which lead to overburdening of the clinician, results in poor accuracy in sleep analysis [10].

With the advent of new technological research techniques in analyzing sleep disorders, newer approaches are introduced in automatic sleep stage classification for analysis of sleep patterns. In these techniques, the sleep experts can easily track

Table 1 Possible sleep stage classification combination

Combination of different possibilities	Sleep stages
Five-state	W, N1, N2, N3 and REM
Four-state	W, N1, (N2 + N3) and REM
Three-state	W, (N1 + N2 + N3) and REM
Two-state	W and sleep (N1 + N2 + N3 + N4 + REM)

different sleep stages from on the subject. Here, human errors are reduced due to less interference during sleep staging analysis. In this research work, we have also obtained the automatic sleep stage classification followed by AASM manuals for our experimental work.

The rest of research work is further managed according in this manner: The existing research contribution of the related work is described in Sect. 2. Section 3 introduces the proposed SleepEEG methodology. Section 4 presents obtained subject's details and data preparation from the dual channel of the EEG signal. Section 5 describes the experimental results and discussion about the proposed study. Finally, concluding remarks of this paper are expressed in Sect. 6.

2 Related Work

Lajnef T. et al. designed an automated sleep stage scoring analysis and he has considered EEG, EOG and EMG signals for their experimental work. Both frequency and time domain features have extracted from input signals. He has employed sequential feature selection techniques for selecting the best combination of features for classifiers. For classification, here, the author has obtained DSVM techniques [11].

Silveira et al. have considered brain signals through single electrodes and obtained features as skewness, kurtosis and variance. Here, author deal with DWT features of the input channel. For classification techniques, here, random forest classifier has employed on the classification of different sleep stages [12].

In [13], here, the author has extracted six channels from EEG signals and has extracted linear features. He has employed the random forest classifier and SVM classifier for classification on different sleep stages.

Fraiwan et al. have proposed a study on sleep stage classification on a newborn baby. He has extracted the time-frequency features which have included continuous wavelet transform techniques for identifying different sleep transition stages [14].

In [15–17], the authors have dealt with EEG signal and obtained both time and frequency related properties from input channel for their experimental work and used some conventional classifiers techniques such as the support vector machine (SVM) [18], the random forest [19] and the KNN [20] for classifying the trained model by considering the extracted features to identify the proper transitions in between sleep stages.

Arthur Flexer et al. have also dealt with a single channel of EEG for his contributed research work. Here, the author has classified between light and deep sleep. He has obtained hidden Markov model for classification purposes to detect the different stages of sleep and its transition phases among them [21].

Tsinalis et al. have considered the healthy adult's dataset for their experiment work on sleep stage classification, here, he has extracted time-frequency-based features from considered channel and has reached an overall accuracy of 86% on EEG data [22].

In [23], the contributor has proposed a model for classifying in between two stages wake versus sleep stages. The obtained input signal segmented into different frequency sub-bands. The extracted properties features from different segments are to be applied through random forest classifiers.

Vural et al. have applied principal component analysis for reducing the high dimensionality time-series signals and he has categorized the feature extraction process into two domains such as linear and nonlinear, and he has compared how best those features discriminating the sleep stages [24].

X. Chen et al. have examined different positioned electrode signals for suitability of sleep abnormality during sleep and the placement of electrodes fixed in the body according to 10–20 electrode system for placing and extracting the signals from the human brain. Here, the authors have followed the clinical standard 10–20 electrode placement procedure for recordings the signals from respected channels of EEG [25].

In [26], the author has designed a sleep stage classification model where he has obtained a single channel from EEG and the author was here applied the Gaussian parameters to find the sleep scoring from subjects. Here, the author has reached the overall accuracy of 90.01%.

Hassan A. R. proposed a scheme using bootstrap aggregating for classification and based on EEG signals from two bench-mark public repository such as Sleep-EDF and DREAMS subject and their accuracy was 92.43% for two-state sleep stage classification problem [27].

In [28], the author has adopted PSG signal for a sleep test, in this observation, the author was extracted features from different frequency bands. Here, the author was applied quadratic discriminant analysis for sleep quality analysis for sleep stage classification. Heyat et al. proposed the sleep disorder analysis by acquiring electroencephalographic signal and obtained cyclic alternating pattern (CAP) sleep dataset from PhysioNet repository for these experimental studies. Here, they suggested power spectral density features from input channel through Welch techniques which alternatively helpful for identifying the depth range of wave patterns in different frequencies ranges. For classification purposes, here, author has used decision tree techniques. An accuracy rate of 81.25% is obtained [29].

3 Methodology

Here, first of all describe the detailed layout on our proposed research scheme on automated monitoring sleep abnormalities based on channel-specific, subject gender-specific and features specific.

3.1 Pre-processing

In this study, we have initially applied the pre-processing techniques for removing the muscle artifacts and noises from the raw signals. For pre-processing the signals, we have adopted the Z score normalization was applied. In Eq. (1), we have mentioned the Z score normalization technique.

$$V' = {(V - \bar{A})}/{\sigma A} \tag{1}$$

where V' is presents the new value of data after normalization operation performed and similarly V contains the old entry of data. σA is the standard deviation and A represents mean of A. Next to Z score normalization, we have also used the second-order butter worth filtering techniques to remove the muscle artifacts from the raw signal.

3.2 Proposed Architecture

In this research work, we design a system based on binary classification in between wake and sleep stages. In this study, we have considered polysomnography signals such as EEG. Here, we consider two individual subjects with different genders. Here, we have recorded signals from two channels such as F3-A2 and C3-A2. In this study, we have only extracted linear features from the processed signals for further experimental work of this proposed study. Figure 1 illustrates the proposed overview of the sleep stage classification architecture.

As per our research work, we have broadly divided the total architecture into five steps. The first step is to be recorded channel signal values according to 10–20 electrode placements from EEG signals of two individual subjects with different sex. Next to signal acquisition pre-processing step obtained for cleaning the noises and irrelevant artifacts. In the third step, we have extracted features from noise-free signals. The recorded signals are characterized with related to properties of time and frequency. Next to extraction of features, the proposed selection techniques select the best feature combinations for classification.

Next to feature selection, we apply the different classifier techniques for classification of sleep stages by considering the selected features. In this study, we also

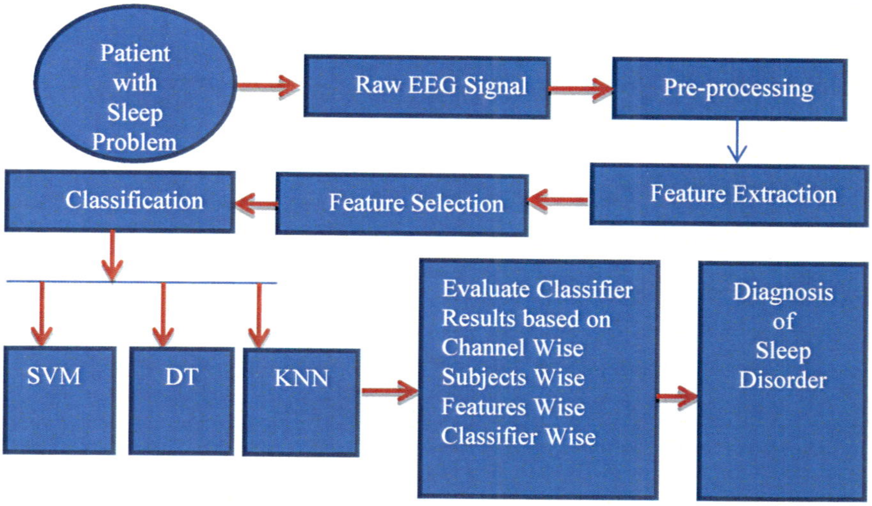

Fig. 1 Workflow of the proposed research work

measured the different index metric considered for this study. We will discuss the different comparison results which are obtained from different classifiers of this proposed study and as per comparison results found from channel and subject wise. Finally, as per sleep scoring achieved, it will make decide what type of treatment requires for proper diagnosis of sleep disorder.

Here, we presents the full view of working architecture and discuss the individual sections in detail. After that we computed performances from different obtained classification techniques in this proposed study. In this study, we also measured the different index metrics considered for this study. We will discuss the different comparison results which are obtained from different classifiers of this proposed study and as per comparison results found from channel and subject wise. Finally, as per sleep scoring achieved, it will make decide what type of treatment requires proper diagnosis of sleep disorder.

In this study, we also describe the complete research work with state chart diagram representation in Fig. 2, where we explained subjects considered in this research work. Besides here also, we mentioned the selected electrode for channel acquisition and features type. In this diagram also, we categorize sleep states classification and the manuals followed for the whole process of sleep scoring to discriminate in between sleep stages.

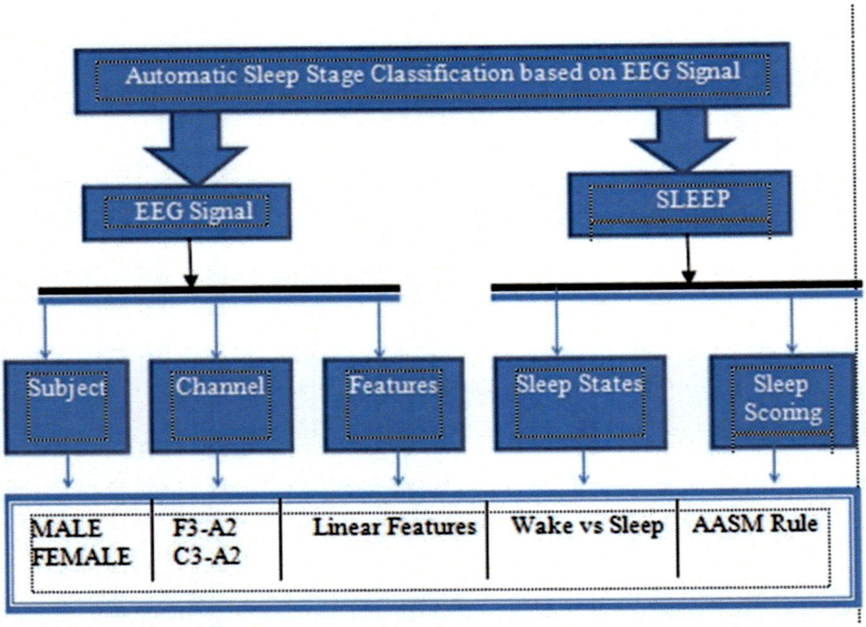

Fig. 2 State chart diagram for two-state sleep stage classification based on EEG

4 Experimental Dataset

We have obtained the subjects information from one of the public sleep data repository called as ISRUC-Sleep. This dataset was derived from the ISRUC-Sleep database; it is publicly available online for researchers who are research in sleep disorder [22]. This whole dataset information was recorded by the sleep experts in the Hospital of Coimbra University (CHUC) in Portugal. This dataset contains 100 subject information; basically, the subjects are in the adult category, including both healthy and with some effected sleep problem. Out of 100 human adults, 53 males, 42 female subjects are there and the rest of the 5 subject's sex is not specified in the database. Data collection was taken from subjects around 8–9 h a full night for individual subjects. This dataset collected signals from 11 electrodes that have placed in the subject's different parts of the body and those electrodes are extracted signals like EEG, EMG, ECG and EOG with sampling rate 200 Hz. In this dataset, the sleep stages are annotated based on the AASM rules.

In this experimental work, we considered only dual channels recorded data such as F3-A2 and C3-A2 of EEG. Here, we have considered only two subjects [Subject No-18 (Male) and Subject No-5 (Female)] with different gender for our experimental work. For our research work for sleep stage classification, we have followed the standard AASM manuals and the concerned EEG recordings and their annotations done through sleep experts. Here, we have extracted 750 epochs with 6000 samples from each subject for both the channels such as F3-A2 and C3-A2.

5 Experimental Results and Discussion

In this research work, we are classifying in between two stages wake versus sleep. For this proposed study, here, we combined NREM and REM stage into one stage called the sleep stage. In this experiment, we considered one male subject and one female subject for identifying which channel is more effective to accurately discriminate the two-state sleep stages. In our proposed method, we have extracted 38 linear features from concerned input channels. The extracted features list is described in Table 2.

Next to the extraction of features from respective channels, we represent the feature selection techniques for finding the best combination of features for the classification task. In this proposed study, we have used online streaming feature selection techniques for selecting suitable features for the classification tasks. Table 3 mentioned the selected best combination of features for classification phase.

Next to feature selection, here, we have used some conventional machine learning classifiers used that are SVM, DT and KNN. The main important work of this research work is to be identifying the sleep transition states between wake states and sleep states. Besides, we also observe which gender subject has generally more inclination toward sleep diseases. We have observed that sometimes gender-specific results differences found as per different state of the artwork done earlier researchers, but it

Table 2 Features used in this proposed study

Label	Short description (frequency domain)	Label	Short description (time domain)
F1	Power	F25	Signal activity
F2, F3, F4, F5	Band power in δ, θ, α, β sub-band	F26 F27	Signal complexity Signal mobility
F6, F7, F8, F9	Relative spectral power in δ, θ, α, β sub-band	F28 F29	Mean Maximum
F10, F11, F12, F13, F14, F15	Power ratio factor for different frequency sub-bands	F30 F31	Minimum Standard deviation
F16	Ratio in between summation of $(\theta + \delta)$ and $(\alpha + \beta)$	F32 F33	Median Variance
F17, F18, F19, F20	Center frequency in δ, θ, α, β sub-band	F34 F35 F36	Zero crossing rate 75 Percentile Skewness
F21, F22, F23, F24	Maximum power in δ, θ, α, β sub-band	F37 F38	Kurtosis Energy

Table 3 Feature selected for individual subject with individual channel

Participants name/gender	Channel	Best feature combination	Classifier
Subject-18 Male	F3-A2	F1, F4, F9, F14, F26, F31, F36 F2, F5, F10, F16, F27 F3, F8, F12, F25, F30 (19 Features)	SVM
			DT
			KNN
	C3-A2	F1, F4, F8, F11, F14, F20, F23, F2, F5, F9, F12, F15, F21, F3, F6, F10, F13 (17 Features)	SVM
			DT
			KNN
Subject-05 Female	F3-A2	F1, F2, F3, F8, F9, F10, F14, F22, F23 F28, F31, F32 (12 Features)	SVM
			DT
			KNN
	C3-A2	F5, F11, F12, F14, F15, F16, F27 (7 Features)	SVM
			DT
			KNN

has less impact regarding sleep scoring as per our research work observation. We have considered k-fold cross-validation techniques in our proposed experimental model where we fix the k value as 10. For comparisons with the different adopted classifiers performances in the subject to sleep scoring, we have also used to calculate some evaluation metrics for measuring the overall performances of proposed SleepEEG test. In this study, we have calculated five indicators such as classification accuracy, sensitivity (also known as recall), specificity, precision and F-Score included in this experiment for evaluation of performances among obtained different classification techniques.

For each stage category c, it is (1) $ACc = TP_c + TN_c/(TP_c + TN_c + FP_c + FN_c)$. (2) $SE_c = TP_c/(TP_c + FN_c)$. (3) $SP_c = TN_c/(TN_c + FP_c)$. (4) $PR_c = TP_c/(TP_c + FP_c)$. (5) $F_c = 2 * (PR_c + SE_c/(PR_c + SE_c)$ [30]. For this proposed study, the whole experimental work is carried out through the Intel i7-6700 processor with 24 GB RAM. The version of MATLAB is 2017a on Windows10 OS platform.

In this study, we found that for a male subject we achieved an overall accuracy level for the C3-A2 channel is more than the F3-A2 channel. We reached the overall accuracy of 95.2% through the SVM classifier. In the same for female subjects, we have received the overall accuracy level from the C3-A2 channel. Here, the overall accuracy level reached 93.7% through KNN classifier as per our observation, we found that the C3-A2 channel is the best-identified channel for identifying the sleep diseases. As the subject in case to classifier, both SVM and KNN have more feasible to adoptable for classification of sleep stages. It has found that linear features are more effective with the C3-A2 channel for both the subjects. To confirm the experiment result from different classifiers, we have also computed the five index parameter. It has observed that we have received the high values for sensitivity, precision and F1 scores and low values for specificity for different classifiers for our proposed study and this observation indicates that the sleep staging classification has accurately identified from subjects enrolled for this proposed study.

The proposed study outcome to create some help for sleep experts regarding taking decisions with regards to handle the subject who are suffered any type of sleep-related disorder. The overall classification accuracy achieved through different classifiers for different channels and different subjects is presented from Figs. 3, 4, 5 and 6. From Tables 4, 5, 6 and 7 represent the performances of different index metrics evaluated from individual channels of subjects enrolled for this research work. Figures 7, 8, 9 and 10 presents the graphical representation of overall performances achieved by obtained different evaluation metrics. To measure the effectiveness of the proposed work, obtained number of approaches to make a comparison with existing contribution in different context like channel acquisition, classification techniques, cross-validation techniques, dataset, feature extraction, feature selection techniques, etc. In this study, we have made comparisons the state of the art contribution with the proposed work in terms of sleep classes and classification techniques used in the experiment work. Table 8 presents the comparison results of the proposed SleepEEG test performances with similar related contributed research work results. Table 8 presents the comparison results of the proposed SleepEEG test performances with similar related contributed research work results.

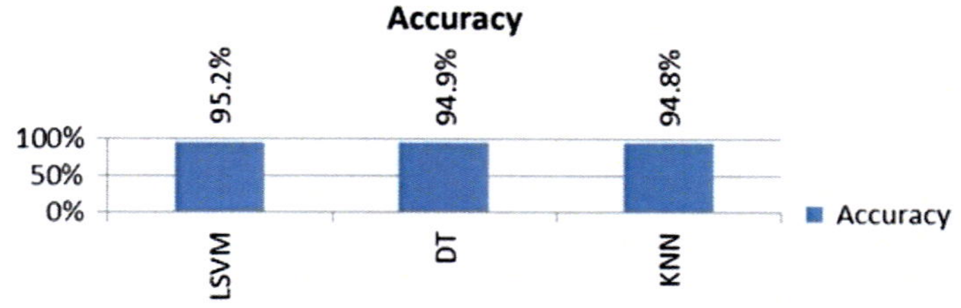

Fig. 3 Overall accuracy of subject-18 (male) for F3-A2 channel

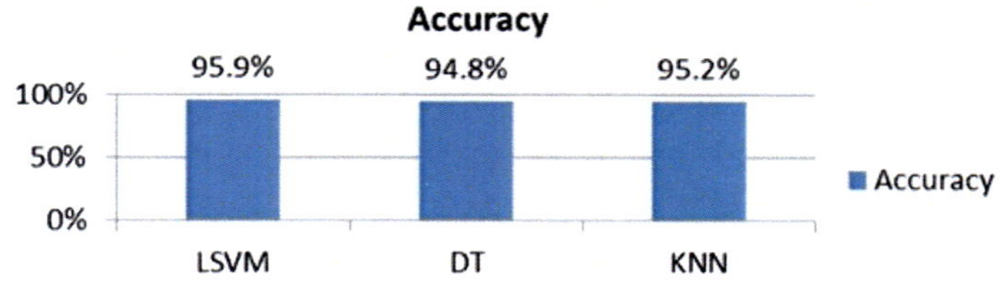

Fig. 4 Overall accuracy of subject-18 (male) for C3-A2 channel

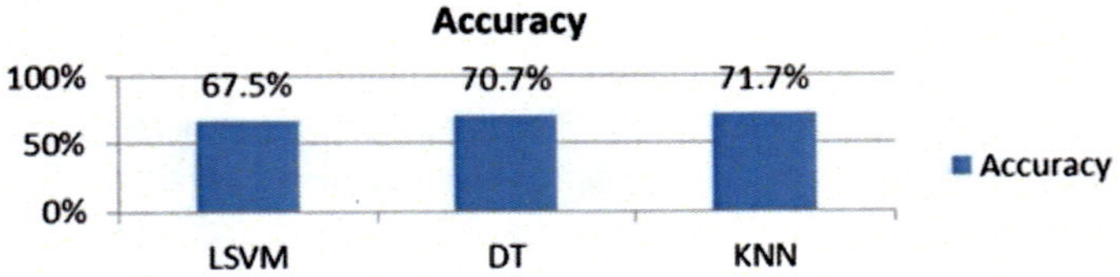

Fig. 5 Overall accuracy of subject-05 (female) for C3-A2 channel

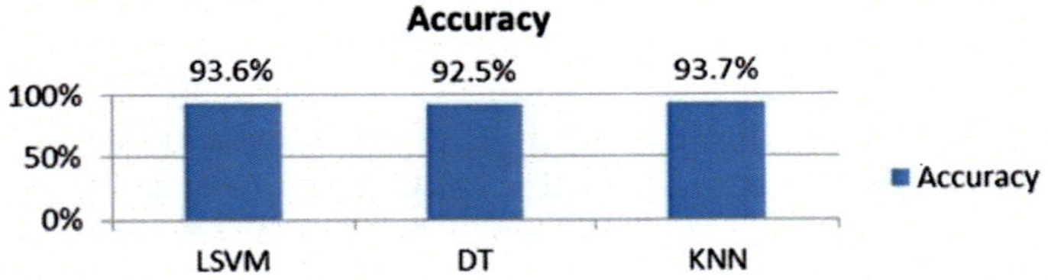

Fig. 6 Overall accuracy of subject-05 (female) for C3-A2 channel

Table 4 Evaluation metrics of subject-18 (male) for channel F3-A2

F3A2	Accuracy (%)	Error rate (%)	Sensitivity (%)	Specificity (%)	Precision (%)	F1-score (%)
LSVM	95.2	4.8	97.5	84.2	96.6	97
DT	94.9	5	98.3	78.9	98.5	98.3
KNN	94.8	5.2	98.5	77.4	77.4	0.968

Table 5 Evaluation metrics of subject-18 (male) for channel C3-A2

C3A2	Accuracy (%)	Error rate (%)	Sensitivity (%)	Specificity (%)	Precision (%)	F1-score (%)
LSVM	95.9	4.1	98	85.7	96.9	97.4
DT	94.8	6.5	97.8	74.8	74.8	96
KNN	95.2	4.8	98.7	78.9	94.3	97.1

Table 6 Evaluation metrics of subject-05 (female) for channel F3-A2

F3A2	Accuracy (%)	Error rate (%)	Sensitivity (%)	Specificity (%)	Precision (%)	F1-score (%)
LSVM	67.5	3.25	100	99.5	67.4	80.5
DT	70.7	2.93	0.928	24.8	71.8	80.9
KNN	71.7	2.82	90.6	32.6	73.5	81.1

Table 7 Evaluation metrics of subject-05 (female) for channel C3-A2

C3A2	Accuracy (%)	Error rate (%)	Sensitivity (%)	Specificity (%)	Precision (%)	F1-score (%)
LSVM	93.6	6.4	94.6	91.4	95.7	95.1
DT	92.5	7.4	94.2	88.9	94.6	94.3
KNN	93.7	6.2	94.6	91.8	95.9	95.2

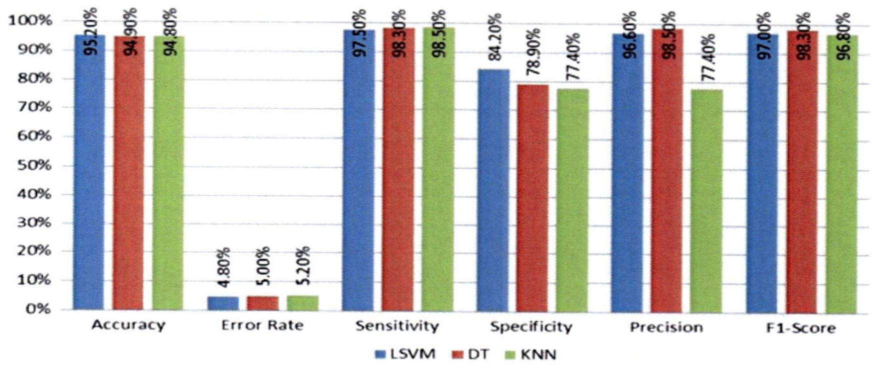

Fig. 7 Subject-18 (male)-F3-A2 (performances of evaluation metrics)

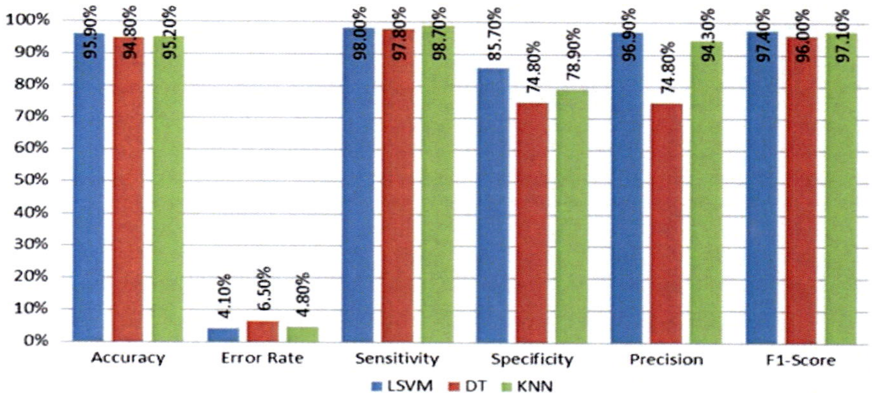

Fig. 8 Subject-18 (male)-C3-A2 (performances of evaluation metrics)

6 Conclusion

Sleep scoring is the first step toward analyzing the sleep quality of subjects. It is the primary approach of any compliant toward sleep diseases. For diagnosing the sleep diseases, primary treatment has to be monitored the different sleep transition

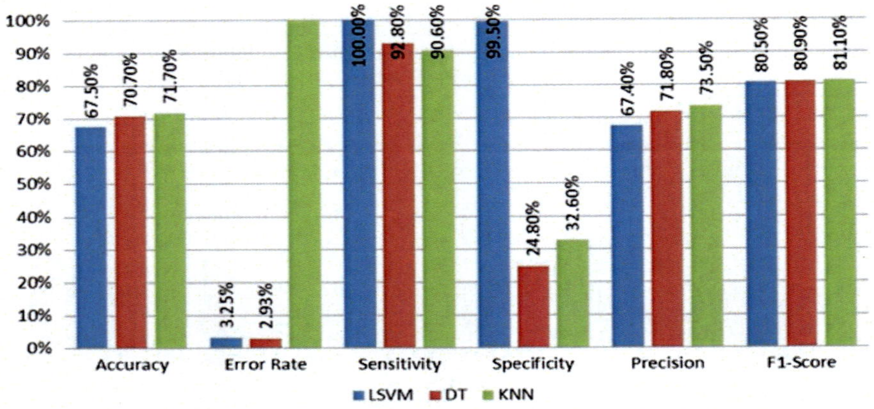

Fig. 9 Subject-05 (male)-F3-A2 (performances of evaluation metrics)

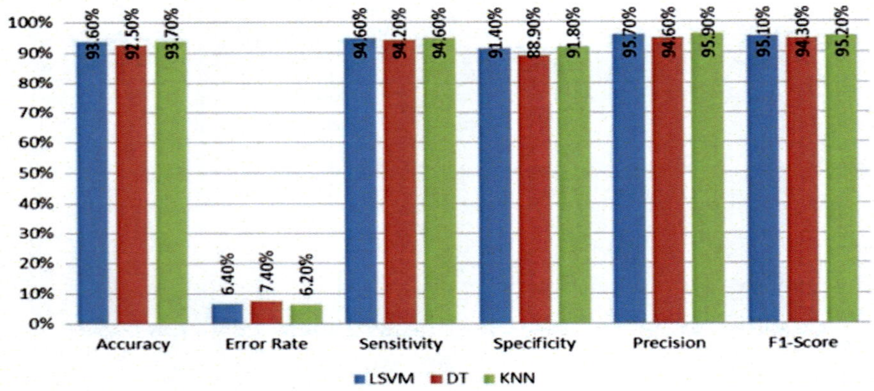

Fig. 10 Subject-05 (male)-C3-A2 (performances of evaluation metrics)

Table 8 Comparison of SleepEEG outcome with the related contributed research works

Authors	Year	Detection	Name of classifier	Signal	Accuracy (%)
Heyat et al. [29]	2019	Sleep disorder	DT	EEG	81.25
Hassan et al. [27]	2017	Sleep disorder	SVM	EEG	92.43
Proposed study	Present	Sleep disorder	SVM	EEG	95.9
			KNN		94.8
			DT		95.2

stages during sleep hours. Therefore, we have used the concept of automatic sleep stage classification techniques in this proposed study. For sleep disorder identification from sleep stages, we have referred automatic sleep stage classification techniques approached, which has used over the years in the field of sleep research. In this proposed study, we have considered the importance of automated sleep staging based on an in scalp-EEG electrodes. For accurate measuring the accuracy level of this proposed model, this has been decided the results from three scenarios: scenario 1 examined sleep score for individual channel to classification among sleep stages, scenario 2 examined automatic score for gender-specific subjects for different channels of brain signals (EEG), scenario 3, here, we examined which combination of linear features has to be appropriate for best sleep stage prediction from dual channel. In scenario 4, we have made a comparison in between different classifiers obtained for this sleep study. Scenario 1 gave the overall accuracy of 95.9% for the C3-A2 channel. According to scenario 2, female subject sleep stage classification has reached the overall accuracy of 95.9% for the C3-A2 channel through SVM classifier, and for KNN, it has to be reached 95.2% and for DT, it has achieved 94.8% overall correctness for identifying the sleep stages. In scenario 3, it has observed that for channel C3-A2 of the subject-18 male category the selected features for classifiers are to be discriminating the transition of different sleep stages accurately. Finally, in scenario 4, it has found that the SVM classifier has to be more effective in terms to classify the sleep stages more correctly. It has reached the overall accuracy 95.9% through SVM classifier. For this experimental study, we have extracted the brain signals from dual channel from two gender-specific subjects, and future application studies will consider a larger group of subjects include the number of channels of EEG, EOG and EMG signals and consider different time frame segmentation for proper diagnosis of sleep stage classification.

References

1. Nugent, C.N., Black, L.I.: Sleep Duration, Quality of Sleep, and Use of Sleep Medication, by Sex and Family Type, 2013–2014. NCHS Data Brief, No. 230. National Center for Health Statistics, Hyattsville, MD (2016)
2. National Sleep Foundation [NFS]: Lack of Sleep is Affecting Americans. https://www.sleepfoundation.org/press-release/lack-sleep-affecting-americans-finds-national-sleep-foundation
3. Sleep Health Foundation [SHF]: https://www.sleepfoundation.org/press-release/lack-sleep-affecting-americans-finds-national-sleep-foundation
4. Sateia, M.: International classification of sleep disorders-third edition. Chest **146**(5), 1387–1394 (2014)
5. Boostani, R., Karimzadeh, F., Nami, M.: A comparative review on sleep stage classification methods in patients and healthy individuals. Comput. Methods Programs Biomed. **140**, 77–91 (2017)
6. Jafari, B., Mohesenin, V.: Polysomnography. Clin. Chest Med. **31**(2), 287–297 (2010)
7. Liang, S.F., Kio, C.E., Hu, Y.H., Y.H. Pan, Wang, Y.H.: Automatic stage scoring of single-channel sleep EEG by using multiscale entropy and autoregressive models. IEEE Trans. Instrum. Meas. **61**(6), 1649–1657 (2012)

8. Sharma, R., Pachori, R.B., Upadhyay, A.: Automatic sleep stages classification based on iterative filtering of electroencephalogram signals. Neural Comput. Appl. **28**(10), 2959–2978 (2017)
9. Berry, R.: Fundamentals of sleep medicine. Philadelphia Elsevier Saunders (2012)
10. Zhu, G., Li, Y., Wen, P.P.: Analysis and classification of sleep stages based on difference visibility graphs from a single-channel EEG signal. IEEE J. Biomed. Health Inform. **18**(6), 1813–1821 (2014)
11. Lajnef, T., Chaibi, S., Ruby, P., Aguera, P.E., Eichenlaub, J.B., Samet, M., Jerbi, K.: Learning machines and sleeping brains: automatic sleep stage classification using decision-tree multi-class support vector machines. J. Neurosci. Methods **250**, 94–105 (2015)
12. Da Silveira, T.L.T., Kozakevicius, A.J., Rodrigues, C.R.: Single-channel EEG sleep stage classification based on a streamlined set of statistical features in wavelet domain. Med. Biol. Eng. Compu. **55**(2), 343–352 (2016)
13. Radha, M., et al.: Comparison of feature and classifier algorithms for online automatic sleep staging based on a single EEG signal. 36th Annual International Conference of the IEEE Engineering in Medicine and Biology Society (EMBC), pp. 1876–1880 (2014)
14. Fraiwan, L., et al.: Time frequency analysis for automated sleep stage identification in full term and preterm neonates. J. Med. Syst. **35**(4), 693–702 (2011)
15. Zafar, R., Dass, SC., Malik, AS.: Electroencephalogram-based decoding cognitive states using convolutional neural network and likelihood ratio based score fusion. PLoS ONE **12**(5) (2017)
16. Zaeri-Amirani, M., et al.: A feature selection method based on Shapley value to false alarm reduction in ICUs a genetic-algorithm approach. In: 2018 40th Annual International Conference of the IEEE Engineering in Medicine and Biology Society (EMBC), pp. 319–323 (2018)
17. Afghah F., Razi A., Soroushmehr R., Ghanbari H., Najarian, K.: Game theoretic approach for systematic feature selection; Application in false alarm detection in intensive care units. Entropy **3**(190) (2018)
18. Koley, B., Dey, D.: An ensemble system for automatic sleep stage classification using single channel EEG signal. Comput. Biol. Med. **42**(12), 1186–1195 (2012)
19. Fraiwan, L., Lweesy, K., Khasawneh, N., Wenz, H., Dickhaus, H.: Automated sleep stage identification system based on time-frequency analysis of a single EEG channel and random forest classifier. Comput. Methods Programs Biomed. **108**(1), 10–19 (2012)
20. Hsu, Y.L., Yang, Y.T., Wang, J.S., Hsu, C,Y.:Automatic sleep stage recurrent neural classifier using energy features of EEG signals. Neuro Comput. 105–114 (2013)
21. Flexera, A., Gruber, G., Dorffner, G.: A reliable probabilistic sleep stager based on a single EEG signal. Artif. Intell. Med. **33**, 199–207 (2005)
22. Tsinalis, O., Matthews, P.M., Guo, Y.: Automatic sleep stage scoring using time-frequency analysis and stacked sparse auto encoders. Ann. Biomed. Eng.
23. Memar, P., Faradji, F.: A novel multi-class EEG-based sleep stage classification system. IEEE Trans. Neural Syst. Rehabil. Eng. **26**(1), 84–95 (2018)
24. Vural, C., Yildiz, M.: Determination of sleep stage separation ability of features extracted from EEG signals using principle component analysis. J. Med. Syst. **34**(1), 83–89 (2008)
25. Chen, X., Liu, A., Peng, H., Ward, R.: A preliminary study of muscular artifact cancellation in single-channel EEG. Sensors **14**(10), 18370–18389 (2014)
26. Hassan, A.R., Bhuiyan, M.I.H.: An automated method for sleep staging from EEG signals using normal inverse Gaussian parameters and adaptive boosting. Neuro Comput. **219**, 76–87 (2017)
27. Hassan, A.R., Bhuiyan, M.I.H.: A decision support system for automatic sleep staging from EEG signals using tunable Q-factor wavelet transform and spectral features. J. Neurosci. Methods **271**, 107–118 (2016)
28. Chapotot, F., Becq, G.: Automated sleep–wake staging combining robust feature extraction, artificial neural network classification, and flexible decision rules. Int. J. Adapt. Control Signal Process. **24**(5), 409–423 (2010)

29. Heyat, M.B.B., Lai, D., Zhang, F.I.K.Y.: Sleep bruxism detection using decision tree method by the combination of C4-P4 and C4-A1 channels of scalp EEG. IEEE Access **1**(1) (2019)
30. Powers, D.M.: Evaluation: from precision, recall and f-measure to roc, informedness, markedness and correlation (2011)

A Tree Multicast Routing Based on Fuzzy Mathematics in Mobile Ad-Hoc Networks

Abu Sufian, Anuradha Banerjee, and Paramartha Dutta

Abstract Nodes in mobile ad-hoc networks are battery powered and moving in arbitrary velocity and direction. So, it is beneficial if nodes has alternative link to successors nodes. The present article proposes a tree multicast protocol where relative mobility of nodes, residual energy, and energy depletion rate along with packet drop rate have considered. Fuzzy mathematics is used to combining these parameters to calculate weight of routes. Simulation results confirm advancement of the proposed protocol over existing state-of-the-art multicast protocols.

Keywords Energy efficient · Fuzzy mathematics · Mobile Ad-hoc networks · Mobility · Multicasting · Routing

1 Introduction

A mobile ad-hoc network (MANET) is an inter-connection of mobile nodes that moves with unpredictable velocity and direction. Here, no centralized administration or infrastructure is there [1, 2]. Therefore, this kind of networks is very effective in emergency-like situations when infrastructure-based networks are unable to work; at the same time, this type of networks is very vulnerable. Several routing strategies have been proposed [3, 4]. Among them, multicast routing is one good way to send data packets to a group of receivers. Many multicast routing protocols have been proposed in MANET [5–10]. Some protocols have considered mobility of nodes

A. Sufian (✉)
University of Gour Banga, Malda, India
e-mail: sufian.csa@gmail.com

A. Banerjee
Kalyani Government Engineering College, Kalyani, India
e-mail: anuradha79bn@gmail.com

P. Dutta
Visva-Bharati University, Santiniketan, India
e-mail: paramartha.dutta@gmail.com

J. K. Mandal et al. (eds.), *Applications of Internet of Things*, Lecture Notes in Networks and Systems 137, https://doi.org/10.1007/978-981-15-6198-6_10

as important parameters, whereas some considered energy efficiency [11, 12]. This current protocol, we called it **TMRF**, is an optimized and fuzzy version of the state-of-the-art protocol WTMR [10]. Here, we considered four important parameters, namely **mobility, residual energy, energy depletion rate**, and **packet drop rate** of participating node; then, using fuzzy mathematics, these are combined to get final decisive parameter.

The rest of the article is organized as follows: In Sect. 2, we have explained the present strategy that is TMRF, and Sect. 3 describes optimum route selection. In Sect. 3.3, computation of weight has explained, whereas Sect. 4 is dedicated to discussion of simulation results, and conclusion of the article is drawn in Sect. 5.

2 TMRF in Details

2.1 *Parameters Used*

It is expected that mobile nodes with maximum residual power shall do better in MANETs, but it may not always true. Energy depletion rate is also very important along with residual energy. Energy depletion rate is a parameter which indicates total residual energy loses in Joules per second. A busy node looses its energy drastically compared to idle node, so it could get down first although it might have more residual energy compared to a idle node which has less residual energy. Therefore, energy depletion rate along with residual energy are considered in TMRF. Mobile nodes in MANET move with unpredicted velocity with arbitrary directions. This is one of the main challenge to maintain connection among nodes in MANETs. So, mobility and frequently routes breaks, as result frequently route establishment phase needs to run, which degrades the performance of a networks. Therefore, this parameter is very important in a MANET, and this is considered in TMRF. Packet drop rate of a node is also considered in TRMF as it is as crucial as some busy node which drops the packet when unable to transfer.

2.2 *Model of the Networks*

Here, the MANET is considered as a graph $G = (V, E)$; here, V is the set of mobile nodes (vertices), and E is the set of links(Edges) among nodes. TMRF modeled subgraph G' from the graph G s.t. $G' = (V', e(s, \alpha(s)))$; $\alpha(s) \in V'$ here V' is the set of multicast groups of nodes each consisting of one sender node and multiple receiver modes. If node n_s is sender, then $\alpha(s)$ is the multicast group. $e(s, \alpha(s))$ is the set of optimum route(paths) from node n_s to each member of $\alpha(s)$ except source node. Each node in the network regularly broadcasts a HELLO messages. All mobile nodes within radio circle of that node then reply with acknowledgment(ACK) message. Formats of HELLO, ACK, and RREQ messages of a node are same as WTMR [10].

If node n_i is downlink neighbor of source node n_s, then node n_i knows the residual energy and energy depletion rate of its own as well as of node n_s by last HELLO message sent by node n_s. Therefore, node n_i can easily calculate expected residual lifetime of the link $n_s \rightarrow n_i$. $ERL(s, i)$ as in Eq. 1. For next node n_j of node n_i, $ERL(i, j)$ was calculated as WTMR. Format of RREQs generated by nodes n_s, n_i, and n_j in different timestamps is same as WTMR.

After arriving all RREQ message at the destination node, the destination node assigns weight to each route and elects one route with the maximum weight, and later packet drop rate is also considered for weight updating. If two routes come with equal weights, then delay will consider to breaks the tie, and even if delay is same, then minimum number of hopes will be considered to break this tie.

3 Optimum Route Selection

3.1 *Estimating Lifetime of a Route*

TMRF first calculates lifetime of each link of a route before estimating lifetime of that route. Link lifetime of a link from node n_i to node n_j at current time is denoted by $ERL(i, j)$, and it is calculated by fuzzy t-norm intersection as Eq. 1.

$$ERL(i, j) = \min(F\_elife(i, j), F\_vlife(i, j)) \tag{1}$$

$elife(i)(\leq MT$, a standard maximum battery life) denotes energy-related link life, and $F\_elife(i)$ is the corresponding fuzzy counterpart; these are calculated using Eqs. 2 and 3, respectively. $F\_elife(i, j)$ is found by fuzzy t-norm intersection between $F\_elife(i)$ and $F\_elife(j)$ as in Eq. 4. Similarly, $vlife$ ($\leq$ 10 Hours, a standard maximum connected time) is for velocity- or mobility-related link life, and F_vlife is the corresponding fuzzy counterpart; these are calculated using Eqs. 7 and 8, respectively.

$$elife(i) = \frac{res\_eng(i) - \{max\_eng(i) \times 0.4\}}{depl\_eng(i)} \tag{2}$$

$$F\_elife(i) = \frac{elife(i)}{MT} \tag{3}$$

$$F\_elife(i, j) = \min(F\_elife(i), F\_elife(j)) \tag{4}$$

Supposed n number of ACK packets comes from node n_j at node n_i, here $p\_trans(i)$ is for the maximum transmission power of node n_i, whereas $distt_l(i, j)$ denotes the distance between nodes n_i and n_j. Let $p\_recv(j, l)$ is the signal power of l-th

$(1 \leq l \leq n)$ ACK packet, and time difference between two successive ACK packet is tme. As per Frii's transmission formula for communication among antennas, $distt_l(i, j)$ is calculated by Eq. 5.

$$distt_l(i, j) = \sqrt[m]{\frac{p\_trans(i) \times K}{p\_recv(j, l)}} \tag{5}$$

Here, K is constant,, and value of m is 2 or 3 depending upon medium. For $2 \leq l \leq n$, if $distt_l(i, j) < distt_{l-1}(i, j)$, then the relative velocity(mobility), $rmv(i, j)$ between nodes n_i and n_j is given by Eq. 6.

$$rmv(i, j) = \sum_{l=0}^{n} \frac{distt_l(i, j) - distt_{l-1}(i, j)}{tme \times n \times rad(i)} \tag{6}$$

Here, $rad(i)$ is the radio range of node n_i. For link $n_i \rightarrow n_j$, if $rmv(i, j) < 0.001$ KM, then $vlife$ is assumed to be 10 h; otherwise, it is estimated by Eq. 7. The node n_j will be out of the radio range of node n_i if it covers at least distance $(rad(i) - cdt(i, j))$,and here, $cdt(i, j)$ is the current distance between nodes n_i and n_j. Therefore, $vlife$ is calculated by Eq. 7.

$$vlife(i, j) = \frac{rad(i) - cdt(i, j)}{rmv(i, j)} \tag{7}$$

$$F\_vlife(i, j) = \frac{vlife(i, j)}{10} \tag{8}$$

Supposed R is one such type route as:
$n_s = n_i \rightarrow n_{i+1} \rightarrow n_{i+2} \rightarrow \ldots \rightarrow n_{i+k} = n_d$. Therefore,

$$minlife(R) = \min\{ERL(i, i + 1), \ \ldots, ERL(i + k - 1, i + k)\} \tag{9}$$

3.2 Estimating Packet Drop Rate

Number of data packets dropped at node n_i is denoted by $PcktDrop(i)$, and it can be easily calculated by Eq. 10.

$$PcktDrop(i) = PcktArr(i) - PcktDept(i) \tag{10}$$

Suppose, $InvPcktDrop(i)$ is the inverse of packet drop rate of node n_i which can take values between 0 and 1 (0 means all packet are drops, and 1 mean no drops), and it can be calculated as in Eq. 11.

$$InvPcktDrop(i) = 1 - \frac{PcktDrop(i)}{PcktArr(i)} \tag{11}$$

Therefore, the packet drop rate of a route R (mentioned in Sect. 3.1) can be estimated by as in Eq. 12.

$$PcktDrop(R) = \min(InvPcktDrop(1), InvPcktDrop(2), .., InvPcktDrop(i), ..) \tag{12}$$

3.3 *Computation of Decisive Weight*

Initially, *minlife* of routes are used to elect some routes for communication initiating, and later packet drop rate is also considered for weight updating. Two calculated parameters, namely $minlife(R)$ and $PcktDrop(R)$ are combined fuzzy max-product composition as in Eq. 13 to get final decisive parameter of a route R.

$$W(R) = \max[minlife(R) * PcktDrop(R)] \tag{13}$$

4 Simulation Results

4.1 *Simulation Environment*

TMRF was implemented in NS-2 [13], and results compare with state-of-the-art protocols ODMRP [14], MAODV [15], and EEMR [16]. Comparison done in terms of: packet delivery ratio, end-to-end delay, multicast route lifetime, and control message overhead. These are measured w.r.t. a number of nodes, number of senders, and mobility of nodes.

Measured number of nodes are 20, 40, 60, 80, and 100. Network sized 1000 $\times$ 1000 m^2. Mobility model used Random Waypoint [17]. Mobility of nodes, i.e., velocity, could be: 10, 20, 30, 40m and 50 km/h. The number of senders at time is 5 to 20 nodes while group size ranges from 5 to 20 nodes. Broadcast channel capacity is 2 Mbps. MAC standard is IEEE 802.11g. Traffic rate is 20 packets per seconds. Packet size is 512 bytes. Maximum queue size of each node is 100 packets. Radio range varies from 50 m to 300 m. Result comparisons are shown in Figs. 1, 2, 3, 4, 5, 6, 7, 8, 9, 10, 11, and 12.

4.2 *Experimental Results*

Packet Delivery Ratio: Packet delivery ratio is a measurement of successfully delivered data packets to group of destinations w.r.t the number of data packets sent to this

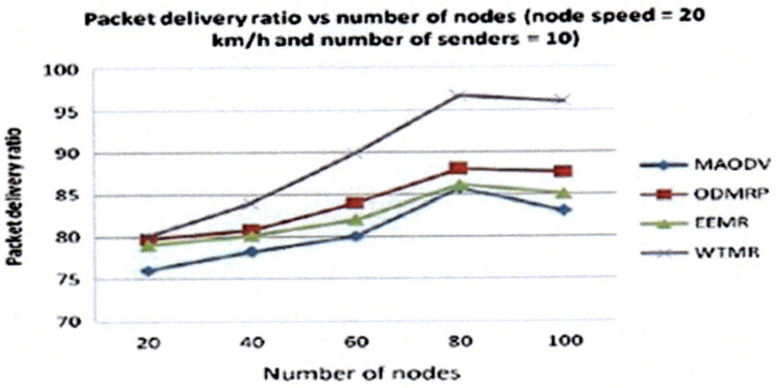

Fig. 1 Packet delivery ratio versus number of nodes

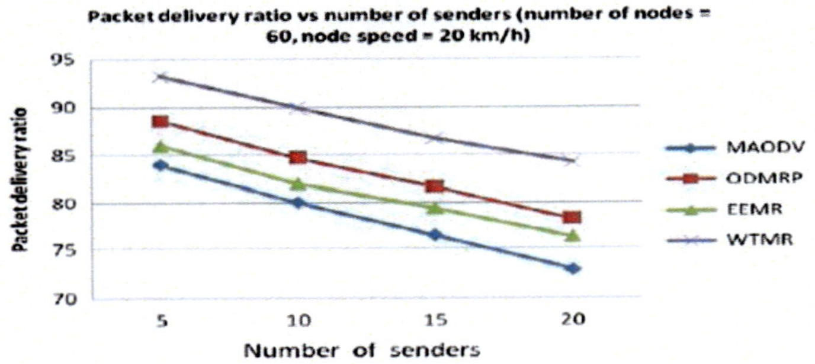

Fig. 2 Packet delivery ratio versus number of senders

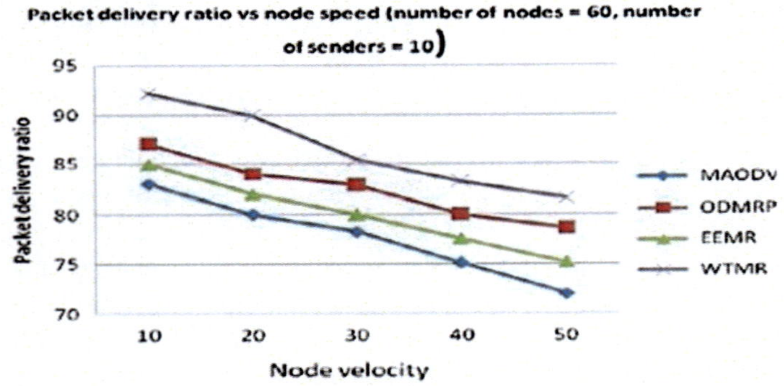

Fig. 3 Packet delivery ratio versus velocity of nodes

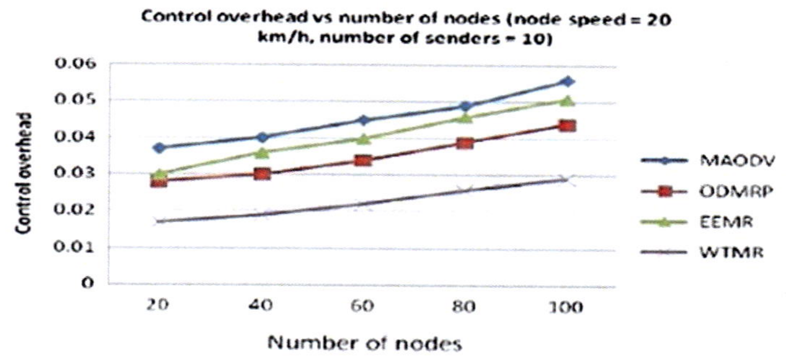

Fig. 4 Control message overhead versus number of nodes

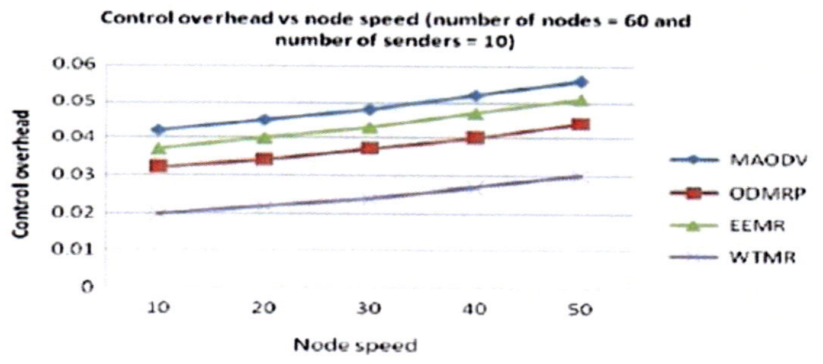

Fig. 5 Control message overhead versus velocity node

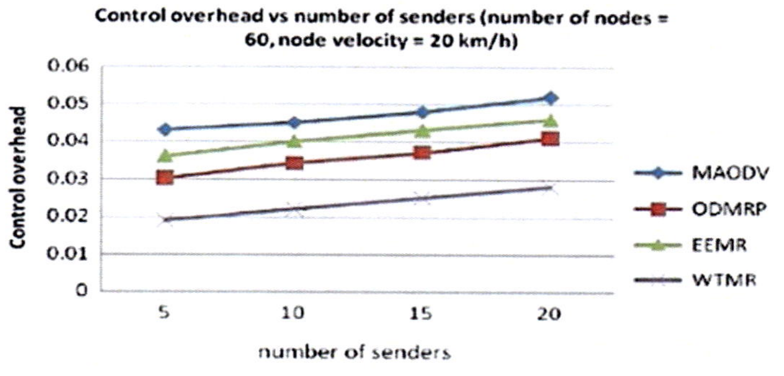

Fig. 6 Control message overhead versus number of senders

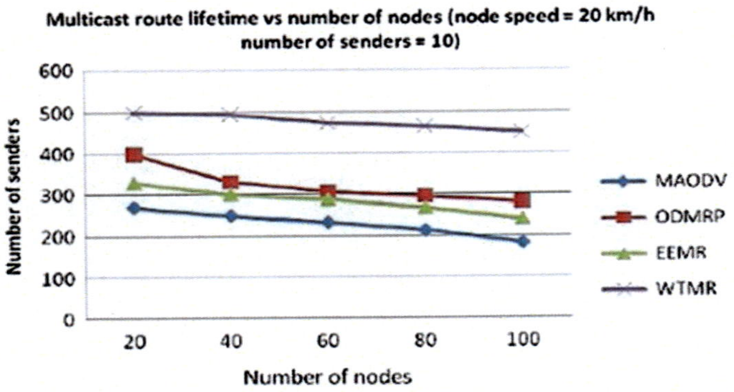

Fig. 7 Multicast route lifetime versus number of nodes

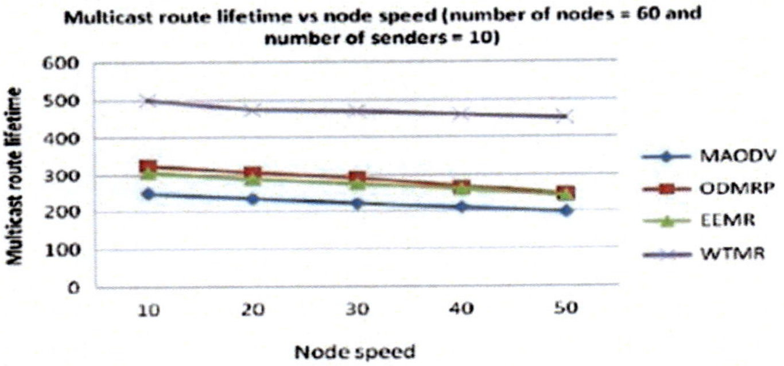

Fig. 8 Multicast route lifetime versus velocity of nodes

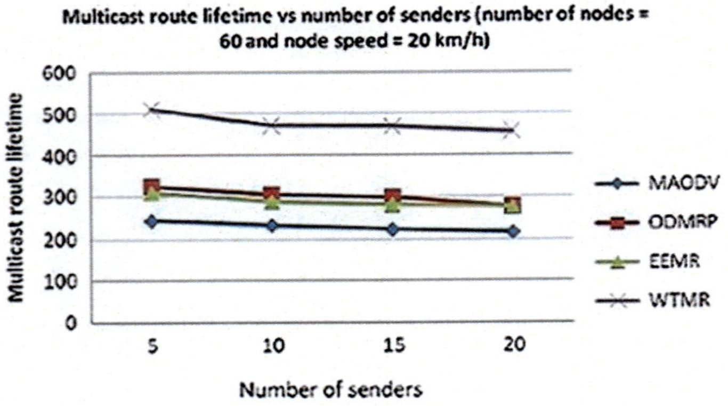

Fig. 9 Multicast route lifetime versus number of senders

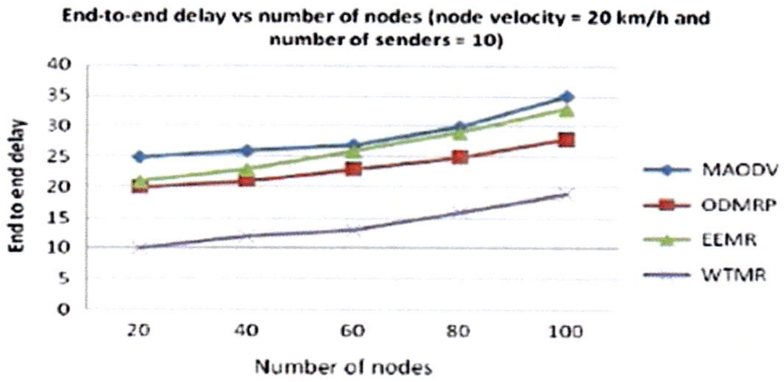

Fig. 10 End-to-end delay versus number of nodes

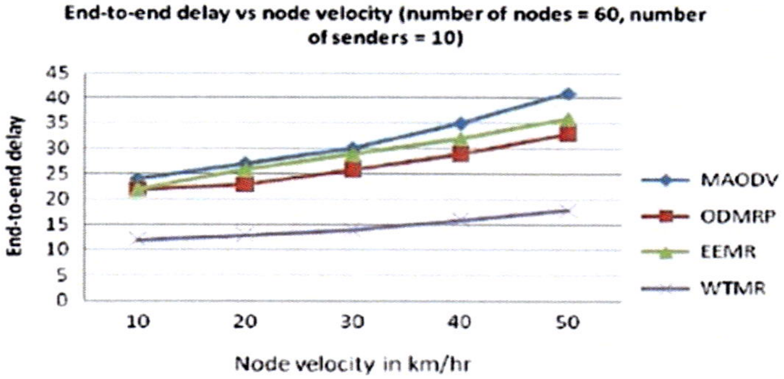

Fig. 11 End-to-end delay versus velocity of node

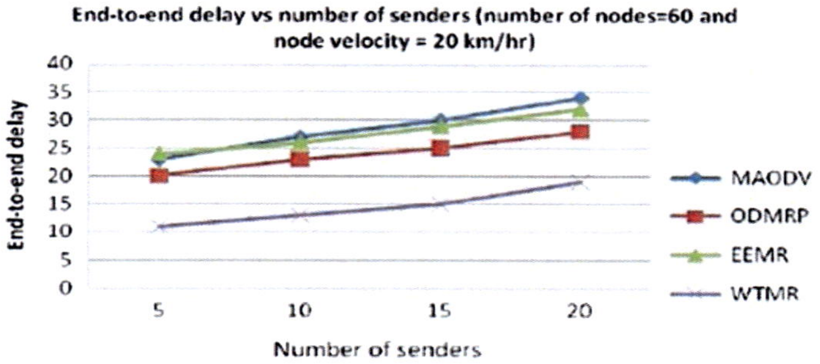

Fig. 12 End-to-end delay versus number of senders

group. It was shown in Fig. 1 w.r.t the number of nodes in the network. Compared to other state-of-the-art multicast routings: MAODV, ODMRP, and EEMR, this TMRF gives better packet delivery ratio. TMRF has considered lifetime of route, although EEMR considers energy efficiency, but as TMRF explain high residual energy may not produce long lifetime. In addition, TMRF has given priority to routes which connect multiple destinations for multicast with comparably stable. As nodes with high residual lifetime with alternative paths are expected to survive more, as a result, TMRF successfully delivers more data packets to destinations compared to others protocols which is shown in Fig. 2. Figure 3 is showing the packet delivery ratio with respect to mobility of nodes. If mobility of nodes increases, then frequently route breakages could happen as well as new link options shall come. As TMRF has less control overhead, collision and contention result better packet delivery ratio. In Figs. 2 and 3, packet delivery ratio decreases of four protocols with an increase in number of senders and mobility of nodes. But in Fig. 1, initially packet delivery ratio increases as number of links increase, and after that, it starts decreases.

Control Message Overhead: Extra control message is a burden to re-establish connections. As in TMRF, lifetime of routes is long so produce less control message overhead, which has shown in Figs. 4, 5, and 6. As expected, control message overhead raises with an increases of mobility of nodes and senders.

Multicast Route Lifetime: This is the parameter for which other parameters of TMRF also produce better results. Unlike the other three state-of-the-art protocols, TMRF directly favor route lifetime which is different from classical energy efficiency routing such as EEMR. In TMRF, alternative routes increase the lifetime of connections, the results can be seen in Figs. 7, 8 and 9. For an increasing of nodes and senders, energy depletion rates increases, as result lifetime also reduces.

End-to-end Delay: End-to-end delay is the time duration from initiating first RREQ to delivering last data packet from sender to multicast receivers. TMRF saves routes re-establishing time by decreasing a number of sessions of route re-discovery, it also reduces control overhead, message collision and contention, and re-sending of data packets. Improvements of TMRF over the others three are shown in Figs. 10, 11, and 12.

5 Conclusion

The TMRF is a multicast protocol which consider four main parameters of MANETs, which are residual energy, energy depletion rates, mobility of nodes, and packet drop rate. By considering this parameters, TMRF calculates weight of each path using fuzzy mathematics and selects best of theme to deliver data packets from source node to a group of receiver's nodes. TMRF gives better performance in simulation results in terms of packets delivery ratios, control overhead, lifetime, and end-to-end delay.

References

1. Chlamtac, I., Conti, M., Liu, J.: Mobile ad hoc networking: imperatives and challenges. Ad Hoc Networks **1**, 13–64 (2003)
2. Corson, S., Macker, J.: Mobile ad hoc networking (manet): routing protocol performance issues and evaluation considerations. https://tools.ietf.org/html/rfc2501.html (1999)
3. Roy, A., Deb, T.: Performance comparison of routing protocols in mobile ad hoc networks. In: Proceedings of the International Conference on Computing and Communication Systems, pp. 33–48. Springer, Berlin (2018)
4. Banerjee, A., Dutta, P., Sufian, A.: Fuzzy-controlled energy-efficient single hop clustering scheme with (FESC) in ad hoc networks. Int. J. Inf. Technol. **10**(3), 313–327 (2018). https://doi.org/10.1007/s41870-018-0133-0
5. Lee, S.J., Su, W.W.Y., Hsu, J., Gerla, M., Bagrodia, R.L.: A performance comparison study of ad hoc wireless multicast protocols. In: Proceedings IEEE INFOCOM 2000. Conference on Computer Communications. Nineteenth Annual Joint Conference of the IEEE Computer and Communications Societies (Cat. No.00CH37064) **2**, 565–574 (2000)
6. de Morais Cordeiro, C., Gossain, H., Agrawal, D.: Multicast over wireless mobile ad hoc networks: present and future directions. IEEE Network **17**, 52–59 (2003)
7. Obraczka, K., Tsudik, G.: Multicast routing issues in ad hoc networks. In: Proceeding of IEEE International Conference on Universal Personal Communication (ICUPC'98)
8. Soni, S.K., Aseri, T.C.: A review of current multicast routing protocol of mobile ad hoc network. In: Proceeding of Second International Conference on Computer Modeling and Simulation ICCMS' 10, vol. 3, pp. 207–211 (2010)
9. Bin Wang, S.K.S.G.: S-remit: a distributed algorithm for source-based energy efficient multicasting in wireless ad hoc networks. In: Proceeding of GLOBECOM 2003, vol. 6, pp. 3519 – 3524 (2003)
10. Sufian, A., Banerjee, A., Dutta, P.: Energy and velocity based tree multicast routing in mobile ad-hoc networks. Wireless Personal Commun. **107**(4), 2191–2209 (2019). https://doi.org/10.1007/s11277-019-06378-y
11. Das, S.K., Yadav, A.K., Tripathi, S.: IE2M: design of intellectual energy efficient multicast routing protocol for ad-hoc network. Peer-to-Peer Networking Appl. **10**(3), 670–687 (2017)
12. Yadav, A.K., Das, S.K., Tripathi, S.: EFMMRP: design of efficient fuzzy based multi-constraint multicast routing protocol for wireless ad-hoc network. Comput. Networks **118**, 15–23 (2017)
13. Issariyakul, T., Hossain, E.: Introduction to Network Simulator NS2, 1st edn. Springer Publishing Company, Berlin (Incorporated) (2010)
14. Lee, S.J., Gerla, M., Chiang, C.C.: On-demand multicast routing protocol. In: Proceeding of IEEE WCNC'99, pp. 1298–1304 (1999)
15. Zhong, M., Fu, V., Jia, X.: Maodv multicast routing protocol based on node mobility prediction. In: Proceeding of International Conference on E-Business and E-Government (ICEE) (2011)
16. Tiwari, V.K., Malviyal, A.K.: An energy-efficient multicast routing (EEMR) protocol in manet. Int. J. Eng. Comput. Sci. **5** (2016)
17. Bettstetter, C., Resta, G., Santi, P.: The node distribution of the random waypoint mobility model for wireless ad hoc networks. IEEE Trans. Mob. Comput. **2**(3), 257–269 (2003). https://doi.org/10.1109/TMC.2003.1233531

Smart Irrigation System Using Internet of Things

Madhurima Bhattacharya, Alak Roy, and Jayanta Pal

Abstract As agriculture is the backbone of Indian economy, it deserves to be modernized. To overcome backwardness of traditional methods of agriculture and to enhance the crop production, to avoid the risk of damaging crops, and to do efficient use of water resources, the latest technology of Internet of things (IoT) is playing a crucial role nowadays. So, this paper "smart irrigation system" is proposed where the soil sensor is used to collect large number of real-time data from the agricultural fields. The sensors interact with each other through Internet connection. The data collected from the sensors sent to the Web server using wireless sensor network. IoT framework analyzes and processes the sensed data. Then, notifications are sent to the farmer's smartphone application periodically. The farmer can track changes in soil moisture. In this way, unnecessary wastage of water can be avoided. This paper discusses the various experiments done in this context and a comparatively low cost system module with sensors and wireless networks for modernized irrigation is represented.

Keywords Smart irrigation · Internet of things · Arduino · Wireless sensor network · Sensors

M. Bhattacharya · A. Roy (✉) · J. Pal
Department of Information Technology, Tripura University, Agartala, Tripura, India
e-mail: alakroy@tripurauniv.in

M. Bhattacharya
e-mail: madhusherlock10894@gmail.com

J. Pal
e-mail: jayantapal@tripurauniv.in

J. K. Mandal et al. (eds.), *Applications of Internet of Things*, Lecture Notes in Networks and Systems 137, https://doi.org/10.1007/978-981-15-6198-6_11

1 Introduction

Farming is the essential need of people as it is the fundamental wellspring of nourishment and it assumes indispensable job in the development of any nation's economy. As indicated by the most recent UN projections, total populace will ascend from 6.8 billion today to 9.1 billion of every 2050, so demand for cereals (for food and animal feed) is projected to reach some 3 billion tonnes by 2050 [1]. Farming is highly unpredictable, because it largely depends on climatic condition such as rainfall, temperature, humidity, and hail, unpredictable events like plants diseases or attack of insect, pests, as well as ups and downs of agricultural markets. Components that influence the crop production to incredible degree are assault of wild creatures and winged creatures when the harvest grows up. The production is declining a direct result of erratic rainfalls of rainstorm, likewise water shortage on summer.

Smart agriculture subject to Internet of things (IoT) will advance to enable cultivators and farmers to diminish waste and improve proficiency in measure of fertilizer, manure, irrigated water, etc. LED lighting, precise control of photoperiod, and soil and environmental sensors can reduce the cost of energy and increase yields. The blend of customary techniques for cultivating with most recent innovations as Internet of things (IoT) and wireless sensor networks (WSN) can prompt modernization of farming [2].

1.1 Internet of Things

The concept of connected device was first introduced since the 1972 but the actual term Internet of things was established by Ashton [3]. It may be depicted as an group of interconnected computing devices consisting of mechanical and digital devices, any items or any living beings. It indicates the capacity to move information over a network without necessity of any human to human or human to computer cooperation. The Internet of things objects consist of sensors, softwares, network connections and necessary electronics and it empowers them to gather and exchange data and make them responsive.

As described in Fig. 1, with regards to interfacing the Internet of things (IoT), there are an apparently overpowering number of alternatives. Cellular, satellite, Wi-Fi, Bluetooth, RFID, NFC, LPWAN and Ethernet are only a portion of the potential approaches to associate a sensor/gadget and within each of these options there can be different providers.

Fig. 1 IoT framework [4]

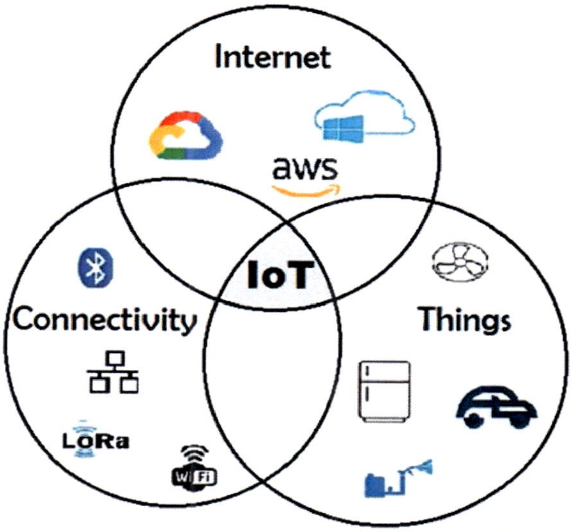

1.2 *Wireless Sensor Networks*

Wireless sensor network (WSN) can be described in Fig. 2 [5] as a distributed network of some devices feature capable of local processing and wireless communication. The devices can communicate the information gathered from a monitored field through wireless links. More specifically, it is a network of small embedded devices called sensors. Sensors are used to collect information from a physical environment. For implementation of wireless communication, industrial areas are necessary because of inaccessibility to remote location, to transmit the information gathered by the sensors and controlling them is not possible every time from a remote location.

The rest of this paper is organized as follows, Sect. 2, reviews the literature related to this work, Sect. 3 presents the architecture of system, module design and working of the referred model. Section 4 discusses the experimental result obtained, and in Sect. 5, there is conclusion and future direction.

2 Literature Survey

The main objective of this project is to design a device which will regulate the usage of water in agricultural field. In the proposed scenario, research has been done to develop an effective automated IoT system by using sensors, Arduino UNO microcontroller board and wireless network. As a reference, we can take the following research paper and architecture of the developed system.

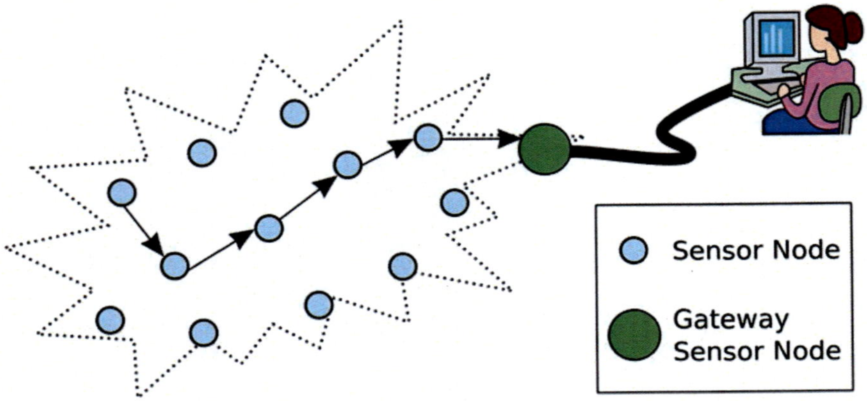

Fig. 2 Wireless sensor network framework

In the proposed work [2], they have discussed the revolution in agriculture industry and elaborated the architecture of IoT system for smart farming. They have discussed how robotics effects in agricultural revolution since 2010. The whole power module of IoT system architecture includes processing module controller processing module memory communication module wireless transmitter-receiver sensing module that is sensors and the interface circuit. Monitored land generates best quantities of data which is stored in the cloud. Small tractors GPS control steering and optimized root planning reduces soil erosion and saving for cost and agricultural drone are applied to monitor this farmland. All the process can be monitored on control center of graphical user interface. Some data has been shared captured by IoT system in smart farming, and the positive results have been explained and analyzed. This paper only discusses the scopes of using IoT in agriculture but does not provide any concise solution or system implementation.

Shadi AlZubi and Bilal Hawashin mentioned in [6] has been introduced Internet of multimedia things (IoMT) that is modification of Internet of things (IoT). IoMT or multimedia wireless sensors network have been utilized in the proposed framework dependent on DIP and MATLAB investigation of the detected multimedia data, and furthermore, an exact hybrid employment work of IoMT approaches with the ideas of machine learning (ML) for irrigation in smart farms. To optimize the irrigation process, this research focused on the smart employment of Internet of multimedia sensors like soil sensor, DHT11 temperature and humidity sensor, light sensor, ultrasonic sensor, rain drop sensor in smart farming. The concepts of image processing work with IoT sensors (IP cameras group I and II) and machine learning methods (WEKA—Waikato Environment for Knowledge Analysis) has been used to make the irrigation decision. Sensor data have been utilized as training data set demonstrating the water requirement of the plants, and AI strategies were utilized in the following stage to locate the ideal choice. The experimental results showed that the use of deep learning proves to be superior in the Internet of multimedia things environment resulting in an optimal irrigation system that reduces both water wastage and manpower.

Josephat Kalezhi and Diana Rwegasira in [7] professed to make a manageable and smart DC microgrid watering system utilizing multi-operator frameworks alongside Internet of things enabled sensors for irrigation. A low cost widely used solar-powered water pumping system has been introduced that uses PV panels, water tanks and water pipes and which can be used to design a loadsheding algorithm to build an irrigation system. An agent-based algorithm regulates energy demand from the PV system and controls irrigation has also been introduced. Data collected from sensor nodes transmitted sequentially over LoRA (Long Range Radio) to a sink node. The automated monitoring using sensors enables the controlled use of limited water resources. As transmission technologies, ZigBee and LoRa have been used for their good communication ranges.

In [8], author proposed a research on smart irrigation system and introduced brief explanation of some application based on IoT to minimize crop loss during harvest or post-harvest with the help of sensors and Raspberry Pi. The proposed system comes up with different smart affordable and profitable module for supervision of soil moisture which used pest sensors, wireless moisture sensor, motor driver, sprinkler, motor alarm, etc., with their usage, shortcomings and advancement. Also, pests intelligent seeds corporation which includes motion and humidity sensors, rodent repellent, dehumidifier starter, camera Raspberry Pi 3, etc., and efficient food corporation of India.

3 Proposed System

In the proposed scenario, in Fig. 3, hardware components used are Arduino board [9], soil sensor, ESP8266 Wi-Fi module, motor driver board, water pump, smartphone of Android operating system, etc., and software platform or language used are Arduino Software, Windows operating system, ThingSpeak IoT platform [10], Android Studio [11], Java, Json, etc. Research has been done to develop an effective IoT system by using sensors, Arduino UNO board and wireless network. It focused on the architecture of Arduino UNO board , ESP 8266 Wi-Fi module, and to build an effective device for smart irrigation system which is more beneficiary than the traditional irrigation system.

In this Sect. 3, proposed scenario has been discussed in subgroups. In Sect. 3.1, the overall architecture of the system has been introduced, Sect. 3.2 discussed module design of the system, Sect. 3.3 has been introduced to represent system methodology and Sect. 3.4 elaborates the working principle.

Fig. 3 Proposed system model

3.1 Architecture

For building up an insightful security gadget dependent on IoT—M2M framework, sensor network and database management are the foundations. Analysts have been creating different IoT-based security gadgets yet a little work is done in agricultural area. For data collection, analysis and transmission the device uses three interfaces. IoT architecture is categorized in three-level architecture and five-level architecture. The working principal of the proposed device based on three-level architecture. In the proposed system, there are three levels of architecture named as—perception layer that is used to differentiate the individual type of sensors; network layer is used for process and transmit the information over network and application layer which is responsible for various practical applications based on users need.

3.2 Module Design

The given Fig. 4 shows the whole architecture of the system. Sensors are spread across the agricultural field which senses and check the moisture content of the soil by soil sensor and through the connectors the data is sent to the Arduino board. The water pumps provided in the field works according to the program uploaded on Arduino board. Sensors and pumps are controlled by the control room. From the control room, data is uploaded into the cloud using ESP8266 Wi-Fi module and after analyzing data, it is sent in the mobile app.

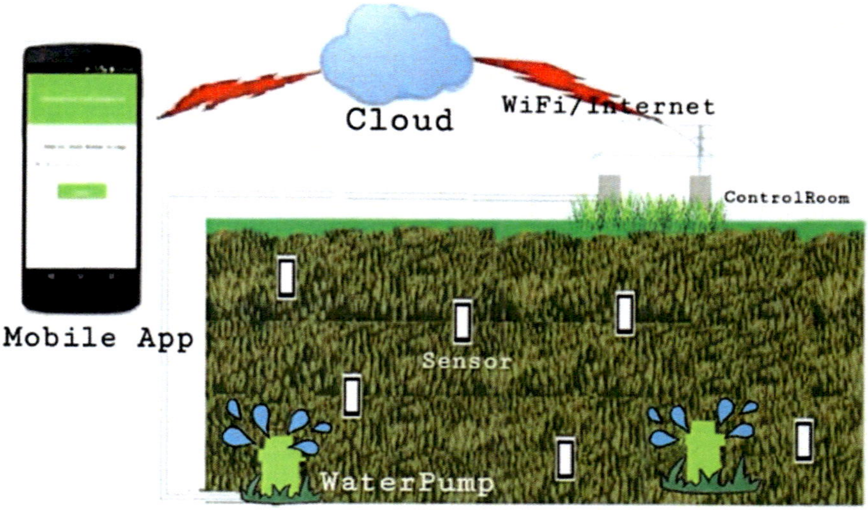

Fig. 4 Architecture of the system

3.3 System Methodology

The YL-38 soil moisture sensor is connected to the Arduino board with connectors. The soil moisture sensor has four pins—VCC, GND and two analog out pins in one side and on the other side there are two pins GND and V+ which are connected to the soil moisture probe. Among the pins A0, GND and VCC pins are connected to the Arduino board. The Arduino board is receiving the sensor data this A0 pin that is connected to the sensor. The Arduino UNO microcontroller board based on ATmega328p has 14 digital input-output pins and 6 analog inputs with a USB connection and a Power Jack on it. ESP 8266 Wi-Fi module consists of 8 pins with patch antenna and a processor (Fig. 5).

RX and TX pin is used for data transmission and reception purpose. The data received by the sensor is uploaded in cloud server with the help of Wi-Fi module. Wi-Fi module uses wireless connection to upload the data into the server. In the proposed system, ThingSpeak IoT platform has been used as cloud server as they are providing free cloud storage. The data from the cloud server is fetched using HTTP POST request and stored in JSON format and extracted by the android app, developed and designed using Android studio. The Android application displays the data to the user. Motor driver board is used for giving extra power to the DC water pump as the power supply is not sufficient for activating the pump.

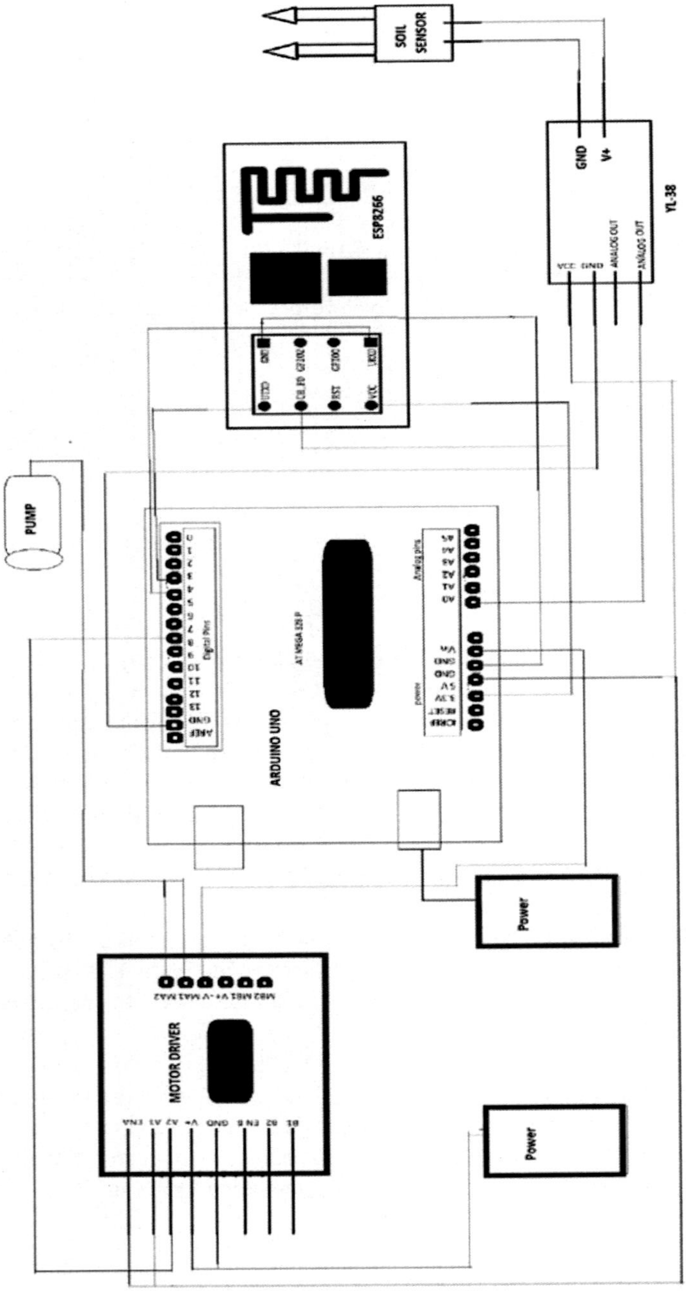

Fig. 5 Circuit diagram of the system

3.4 *Working Principle*

Program run on Arduino board is to fetch the data from the sensor and check the sensor data is greater than or less than the given threshold value to start the water pump is the data is less than the given threshold value the pump will automatically start stop automatically is the sensor data is found higher than the threshold value. Threshold value is given according to the weather condition and soil moisture content of a particular area where the system is implemented. The analyzed data is farther stored in the SQL database provided by the ThingSpeak IoT platform using URL command line tool and library through HTTP protocol. The Android app fetches the data from cloud server on a particular timestamp of the day. The variation of sensor data can be displayed graphically in the app. User can view the data whenever the user wants. Whenever the sensor data will go down below the threshold value user will be notified by an alarm.

4 Experimental Result

Our experimental output, briefly described in Table 1 shows that automated working of water pump and data updating and retrieving operations in cloud server has been done by the proposed system. Program run on Arduino board fetches the data from the sensor and check the sensor data is greater than or less than the given threshold value to start the water pump. If the data is less than the given threshold value the pump will automatically start and stop automatically if the sensor data is found higher than the threshold value. Threshold value is given according to the weather condition and soil moisture content of a particular area where the system is implemented. The analyzed data is stored in the SQL database provided by the ThingSpeak IoT platform using URL command line tool and library through HTTP protocol. The Android app fetches the data from the cloud server on a particular timestamp of the day.

5 Conclusion and Future Direction

The developed system is beneficial for the users and works in a cost-effective manner. It reduces water consumption to a greater extent. The system can be used in green houses and also it will be very useful in areas where water scarcity is a major problem. The harvest efficiency will increment and wastage of yields will be diminished utilizing this water system framework. The created framework is progressively useful and gives increasingly doable outcomes. The smart irrigation system will prove itself as a cost-effective system for optimizing water resources for agricultural production. This project can be extended to a great extent to accelerate the production of crops. Along with the soil moisture temperature humidity also can be detected

Table 1 Experimental result obtained

S. No.	Experiment done	Result
1	Check the sensor by connecting with Arduino board	Sensor data showed in the serial monitor
2	Connect the ESP8266 Wi-Fi module with the existing Wi-Fi network	Serial monitor of Arduino IDE showed User ID and Password of the connected of Wi-Fi network
3	Check the status of the pump when sensor data is less than threshold value	Water pump activated automatically when sensor returned numeric data less than threshold value
4	Check the status of the sensor when sensor data is showing that there is enough water in the soil	Water pump deactivated automatically when sensor returned numeric data greater than threshold value
5	Check if the numeric data of the sensor and Boolean data from the pump are uploading in the intended cloud server provided by ThingSpeak IoT platform	After Logging in into the ThingSpeak Web site, we can view the real-time data of the sensor and the pump which is the graphically represented
6	Check if the user can view the real-time data of the soil sensor and status of the pump through android app	User can see the moisture content of the soil and pump status in the android application installed in his/her smartphone

by the sensors and the whole data set can be uploaded into the cloud and can be further analyzed so that farmer can monitor all the factors related to the growth of crops. Not only that, farmland can be monitored by using cameras and sensors which can protect the farmland from human, rodents, mammals, etc. Further research can improve the functioning of the system and its applicable areas. Internet of things (IoT) has a great possibility to improve our lives.

References

1. @miscFAONewsA69:online, author = , title = FAO-News Article:2050: A third more mouths to feed, howpublished = http://www.fao.org/news/story/en/item/35571/icode/,month=,year=, note=. Accessed on 31 Mar 2020
2. Mat, I., Kassim, M.R.M., Harun, A.N., Yuso, I.M.: Smart agriculture using internet of things. In: 2018 IEEE Conference on Open Systems (ICOS), pp. 54–59. IEEE, New York (2018)
3. Ashton, K., et al.: That internet of things thing. RFID J. **22**(7), 97114 (2009)
4. 3 simple questions for an iot definition [examples] | iot architect. https://www.iot-architect.de/3-simple-questions-for-an-iot-definition. Accessed on 11 Dec 2019

5. Wireless sensor network - wikipedia. https://en.wikipedia.org/wiki/Wirelesssensornetwork. Accessed on 16 Dec 2019
6. AlZubi, S., et al.: An efficient employment of internet of multimedia things in smart and future agriculture. Multimedia Tools Appl. 1–25 (2019)
7. Kalezhi, J., et al.: A DC microgrid smart-irrigation system using internet of things technology. In: 2019 IEEE PES/IAS PowerAfrica. IEEE, New York (2019)
8. Das, R.K., Panda, M., Dash, S.S.: Smart Agriculture System in India Using Internet of Things. Soft Computing in Data Analytics, pp. 247–255. Springer, Singapore (2019)
9. Arduino - software. https://www.arduino.cc/en/Main/Software. Accessed on 16 Dec 2019
10. Iot analytics - thingspeak internet of things. https://thingspeak.com/. Accessed on 16 Dec 2019
11. Download android studio and sdk tools | android developers. https://developer.android.com/studio. Accessed on 16 Dec 2019

Modeling and Analytical Analysis of the Effect of Atmospheric Temperature to the Planktonic Ecosystem in Oceans

Sajib Mandal, M. S. Islam, and M. H. A. Biswas

Abstract In marine ecosystems, plankton is considered as the primary food producer. The growth of plankton depends on the efficiency of saturation carbon dioxide, saturation oxygen, nutrition, temperature of the water, sunlight, saturated or unsaturated toxic chemical, plastic, etc. But the growth of phytoplankton mostly depends on the photosynthetic activity of plankton. On the other hand, the photosynthetic activity varies with different atmospheric temperatures. In this study, we discuss the effect of atmospheric temperature on the plankton in marine ecosystems including the concentration of dissolved oxygen. To investigate the effect of atmospheric temperature, we formulate a mathematical model consists of nonlinear ordinary differential equations considering four dynamical variables as the amount of atmospheric temperature, the density of phytoplankton, the density of zooplankton, and the concentration of dissolved oxygen. After testing the positivity, stability analysis has been performed at different critical points of the proposed model. From numerical simulations, an approximate solution for every dynamical species has been found.

Keywords Atmospheric temperature · Photosynthetic activity · Plankton

---

S. Mandal (✉) · M. S. Islam
Department of Mathematics, Bangabandhu Sheikh Mujibur Rahman Science and Technology University, Gopalganj 8100, Bangladesh
e-mail: sajibmandal1997@gmail.com

M. S. Islam
e-mail: sirajulku@gmail.com

M. H. A. Biswas
Mathematics Discipline, Khulna University, Khulna, Bangladesh
e-mail: mhabiswas@yahoo.com

J. K. Mandal et al. (eds.), *Applications of Internet of Things*, Lecture Notes in Networks and Systems 137, https://doi.org/10.1007/978-981-15-6198-6_12

1 Introduction

Total energy of the marine ecosystem in the ocean is supplied by the plankton population, and the density of plankton greatly depends on photosynthesis. Photosynthetic activity of plankton is a chemical reaction which mostly interacts with temperature [1]. Generally, it starts to increase with the increase of temperature (up to 77 °F or 25 °C) and starts to decrease with high temperature (from 25 °C up to 40 °C). The photosynthetic activity is performed with a very small rate above 40 °C temperature and at a time it will be stopped. Generally, the photosynthesis runs well from 22 to 28 °C. Therefore, this temperature is considered as perfect temperature and 25 °C is considered as the optimum temperature for the photosynthetic activity of phytoplankton [2].

Schabhuttl et al. [3] discussed the statistical analysis of the combined effect of temperature and diversity on phytoplankton's growth considering 15 species of freshwater phytoplankton. Striebel et al. [4] statistically analyzed the difference between the shape of abiotic and biotic, and the response of temperature change to two aspects: permanent raising of average environmental temperature versus tremble disturbance in type of a heat wave. Destania et al. [5] and Khare et al. [6] described the effect of nutrients on plankton through mathematical modeling. Sekerci and Petrovskii [7] described the effect after climate change on plankton–oxygen dynamics. Promrak and Rattanakul [8] described the effect of increasing global temperature on the green lacewings and the life cycles of mealybugs. Besides, some papers [9, 10] described the analytical analysis of the effect of temperature on phytoplankton.

In this study, we proposed a nonlinear mathematical modeling to describe the effect of atmospheric temperature on plankton in marine ecosystems including the concentration of dissolved oxygen. To formulate the model, four dynamical species are considered and other interactions are neglected in this model.

We can easily find out the acuity of photosynthesis of phytoplankton in the ocean with respect to the depth of water from the proposed model. It also helps to acquire knowledge about the relationship between atmospheric temperature and depth from the water surface level, and the corresponding results to the growth of plankton.

2 Model Formulation

To formulate the model of the effect of atmospheric temperature on the plankton of the marine ecosystem, a system of nonlinear differential equations consists of four dynamical species considering the atmospheric temperature (T), the density of phytoplankton (P), the density of zooplankton (Z), the concentration of dissolved oxygen (D). The interrelationship among them can be represented through a diagram.

Figure 1 shows that temperature helps phytoplankton to produce food and phytoplankton serves the energy and oxygen to zooplankton. Thus, they make a balancing marine ecosystem among themselves.

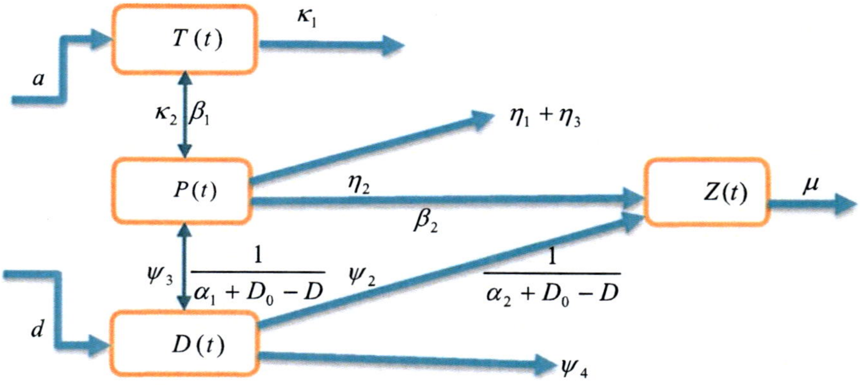

Fig. 1 Schematic diagram of the system representing the interaction among the considered species in marine ecosystem

From the above discussion and according to Fig. 1, the proposed four species ecosystem can be represented by a system of nonlinear ordinary differential equations as:

$$\frac{\mathrm{d}T}{\mathrm{d}t} = a - \kappa_1 T - \kappa_2 PT \tag{1}$$

$$\frac{\mathrm{d}P}{\mathrm{d}t} = \frac{\beta_1 TP}{\alpha_1 + D_0 - D} - \eta_1 P - \eta_2 PZ - \eta_3 P \tag{2}$$

$$\frac{\mathrm{d}Z}{\mathrm{d}t} = \frac{\beta_2 PZ}{\alpha_2 + D_0 - D} - \mu Z \tag{3}$$

$$\frac{\mathrm{d}D}{\mathrm{d}t} = d + \psi_1 PT - \psi_2 DZ - \psi_3 DP - \psi_4 D \tag{4}$$

with initial conditions $T(0) > 0,\ P(0) \geq 0,\ Z(0) \geq 0,\ D(0) \geq 0$.

The brief description of the parameters used in the model is shown in Table 1.

3 Analytical Analysis

In the analytical section, we perform the positivity test of the dynamical variables, stability analysis at equilibrium points, and numerical simulation [11, 12].

3.1 Boundedness of the System

Now, we establish that the system is bounded by using the following lemma.

Lemma 1 *The set* $\Phi = \left\{(T, P, Z) \in \Re_4^+ : 0 \leq T + P + Z \leq \frac{a}{\delta_n}, D \leq \frac{d}{\psi_4}\right\}$ *is a region of attraction for each solution and initially all the variables are positive, and where* $\delta_n = \text{Min}\{\kappa_1, (\eta_1 + \eta_3), \mu\}$.

Proof Let us consider a function $\dot{x}(t) = f(x, t)$, where $x(t) = (T(t), P(t), Z(t))$. If $\delta_n = \text{Min}\{\kappa_1, (\eta_1 + \eta_3), \mu\}$, then we obtain the following inequality:

$$\frac{\mathrm{d}x(t)}{\mathrm{d}t} + \delta_n x(t) \leq a$$

Applying the differential inequalities, we have $0 \leq x(t) \leq \frac{a}{\delta_n}$. Similarly from Eq. (4), we get $0 \leq D(t) \leq \frac{B}{c_1}$, where $c_1 = \psi_2 Z + \psi_3 P + \psi_4$ and $B = d + \psi_1 PT$.

Hence, the solution of the system is bounded in Φ.

3.2 Equilibrium Points

We obtain three equilibrium points of the system (1–4) by setting $\frac{\mathrm{d}T}{\mathrm{d}t} = 0$, $\frac{\mathrm{d}P}{\mathrm{d}t} = 0$, $\frac{\mathrm{d}Z}{\mathrm{d}t} = 0$ and $\frac{\mathrm{d}D}{\mathrm{d}t} = 0$. The equilibrium points are

(i) $E_1(\overline{T}, 0, 0, \overline{D})$, (ii) $E_2(\overline{T}, \overline{P}, 0, \overline{D})$ and (iii) $E_3(\overline{T}, \overline{P}, \overline{Z}, \overline{D})$.

3.3 Stability Analysis

The system of Eqs. (1)–(4) can be represented into Jacobian matrix as

$$J_i = \begin{bmatrix} -\kappa_1 - \kappa_2 P & -\kappa_2 T & 0 & 0 \\ \frac{\beta_1 P}{\alpha_1 + D_0 - D} & \frac{\beta_1 T}{\alpha_1 + D_0 - D} - \eta_1 - \eta_2 Z - \eta_3 & -\eta_2 P & \frac{\beta_1 T P}{(\alpha_1 + D_0 - D)^2} \\ 0 & \frac{\beta_2 Z}{\alpha_2 + D_0 - D} & \frac{\beta_2 P}{\alpha_2 + D_0 - D} - \mu & \frac{\beta_2 P Z}{(\alpha_2 + D_0 - D)^2} \\ \psi_1 P & \psi_1 T - \psi_3 D & -\psi_2 D & -\psi_2 D - \psi_3 P - \psi_4 \end{bmatrix} \tag{5}$$

where $i = 1, 2, 3$

Stability Analysis at E_1. After solving the characteristic equation of (5) at E_1, we get four eigenvalues as

$$\lambda_1 = -\kappa_1,\ \lambda_2 = \frac{a\beta_1/\kappa_1}{\alpha_1 + D_0 - \frac{d}{\psi_4}} - \eta_1 - \eta_3,\ \lambda_3 = -\mu,\ \lambda_4 = -\psi_4$$

Among the four eigenvalues, three of them are negative and one of them may be negative or positive. Then, the equilibrium point E_1 will be stable if $\lambda_2 < 0$.

Stability Analysis at E_2. After solving the characteristic equation of (5) at E_2, we get four eigenvalues as

$$\lambda_1 = -\kappa_1 - \kappa_2\overline{P},\ \lambda_2 = \frac{\beta_1\overline{T}}{\alpha_1 + D_0 - \overline{D}} - \eta_1 - \eta_3,\ \lambda_3 = -\psi_3\overline{P} - \psi_4,$$
$$\lambda_4 = \frac{\beta_2\overline{P}}{\alpha_2 + D_0 - \overline{D}} - \mu$$

where two of them are negative and two of them may be negative or positive. If they are negative, E_2 will be stable, else E_2 will be unstable saddle point.

Stability Analysis at E_3. After solving the characteristic equation of (5) at E_3, we get four eigenvalues as

$$\lambda_1 = -\kappa_1 - \kappa_2\overline{P},\ \lambda_2 = \frac{\beta_1\kappa_1\overline{T}}{(\alpha_1 + D_0 - \overline{D})(\kappa_1 + \kappa_2\overline{P})},\ \lambda_3 = \frac{\beta_2\overline{P}}{\alpha_2 + D_0 - \overline{D}} - \mu,$$
$$\lambda_4 = -\psi_2\overline{Z} - \psi_3\overline{P} - \psi_4$$

where two of them are negative and two of them may be negative or positive. If they are negative, E_3 will be stable, and if they are positive, E_3 will be unstable saddle point.

3.4 Numerical Simulations

Graphical representation through numerical simulation is the most useful task to represent the interactions among the dynamical variables. Here, to check the feasibility of our analysis concerning stability axioms, we use Maple coding. Some numerical computations have been driven by using these coding choosing a set of parameters shown in Table 1. The conditions for the existence of interior equilibrium E_3 are satisfied under these parametric values, and the numerical solutions for each dynamical species are obtained at temperature 25 °C (shown in Fig. 3) given as

$$\overline{T} = 11.53,\ \overline{P} = 0.851,\ \overline{Z} = 1.379,\ \overline{D} = 9.547$$

Table 1 Brief description of the parameters in the model is as follows

Symbol	Meaning	Values
a	Atmospheric temperature of the earth	25 °C
κ_1	Rate of system loss	$0.78\,\text{kl}^{-1}$
κ_2	Absorbing rate of temperature for photosynthesis	$0.300\,\text{kl}^{-1}$
β_1	Proportional constant	$0.50\,\text{day}^{-1}$
α_1	Saturation constant	$0.51\,\text{mg}\,\text{l}^{-1}$
η_1	Natural death rate of phytoplankton	$0.009\,\text{day}^{-1}$
η_2	Predation rate of zooplankton	$0.41\,l\,\text{mg}^{-1}\,\text{day}^{-1}$
η_3	Density of water (muddy and dirty)	$0.01\,\text{mg}^{-1}\,\text{l}^{-1}$
β_2	Proportional constant	$0.33\,\text{day}^{-1}$
α_2	Saturation constant	$0.41\,\text{mg}\,\text{l}^{-1}$
μ	Natural death rate of zooplankton	$0.01\,\text{day}^{-1}$
d	Concentration of dissolved oxygen enters into the system	$24\,\text{mg}\,\text{l}^{-1}\,\text{day}^{-1}$
ψ_1	Producing rate of O_2 by photosynthetic activity	$0.652\,\text{mg}\,\text{l}^{-1}\,\text{day}^{-1}$
ψ_2	Absorbing rate of O_2 by zooplankton for breathing	$0.02\,\text{mg}\,\text{l}^{-1}\,\text{day}^{-1}$
ψ_3	Absorbing rate of O_2 by phytoplankton for respiration	$0.025\,\text{day}^{-1}$
ψ_4	Natural depleting rate	$3\,\text{day}^{-1}$
D_0	Saturation value of dissolved oxygen	$30\,\text{mg}\,\text{l}^{-1}$

Figure 2 represents the effect of lower temperature (18 °C) on photosynthesis and the corresponding effect on zooplankton and oxygen. At temperature 18 °C, the rate of photosynthetic activity of phytoplankton is not optimum and so grows on. On the other hand, the growth rate of zooplankton and oxygen is increasing because of the increasing rate of phytoplankton. Figure 2a shows that the atmospheric temperature decreases proportionally with the depth of ocean measured from the water surface layer.

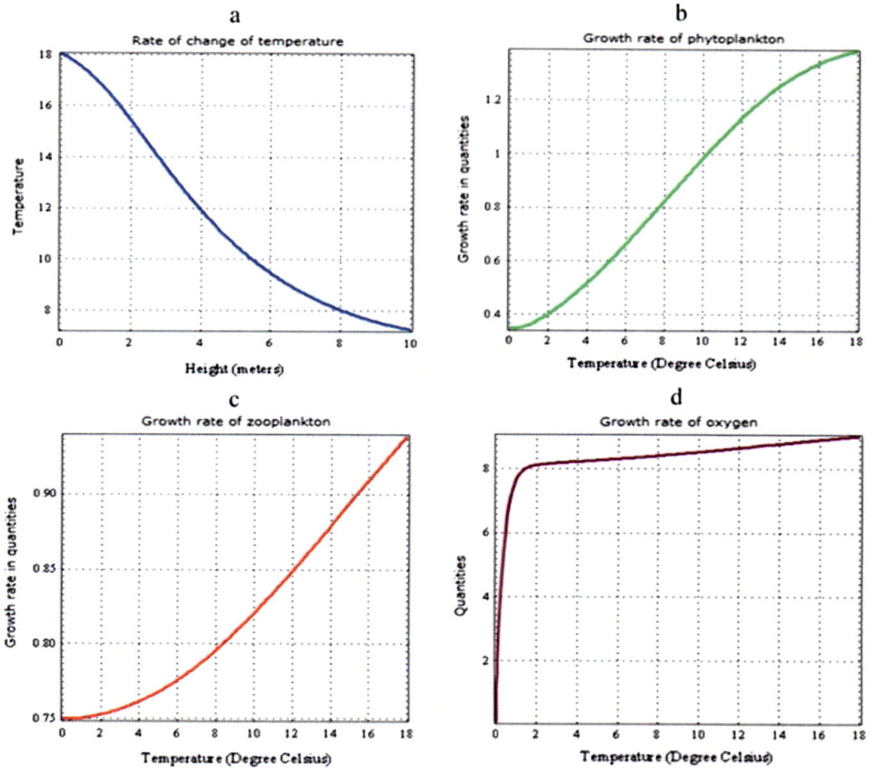

Fig. 2 Effect of temperature on planktonic ecosystem under $a = 18\,°\text{C}$, $\kappa_1 = 0.72\,\text{kl}^{-1}$, and $\kappa_2 = 0.250\,\text{kl}^{-1}$

When the temperature reaches to the optimum temperature (25 °C), the rate of photosynthetic activity is maximized as shown in Fig. 3b. So at that temperature, the rate of photosynthesis remains constant with time. As a result, the growth of zooplankton and oxygen becomes maximized with the highest growth rate as shown in Fig. 3c, d, respectively. We notice that the absorbing rate of temperature by phytoplankton and system loss of temperature are proportionally changed with the atmospheric temperature.

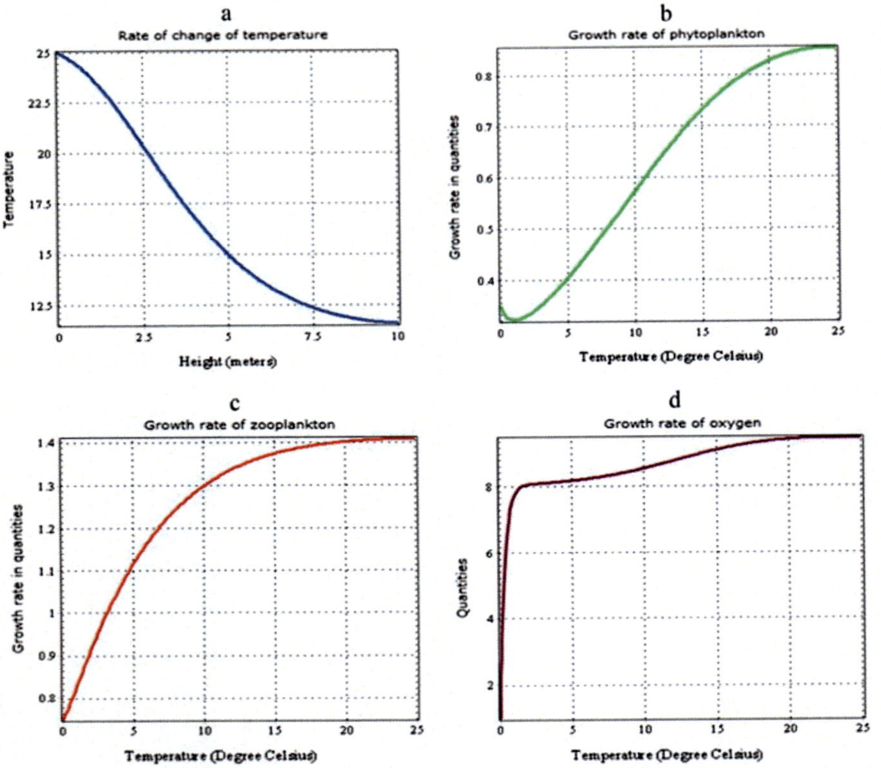

Fig. 3 Effect of temperature on planktonic ecosystem under $a = 25\,°\text{C}$, $\kappa_1 = 0.78\,\text{kl}^{-1}$, and $\kappa_2 = 0.300\,\text{kl}^{-1}$

Figure 4 shows the effect of over optimal temperature to the system. When the temperature crosses the optimal state, the photosynthesis starts to decrease. With the decreasing rate of phytoplankton, the growth rate of zooplankton and oxygen will be decreased proportionally as shown in Fig. 4c, d.

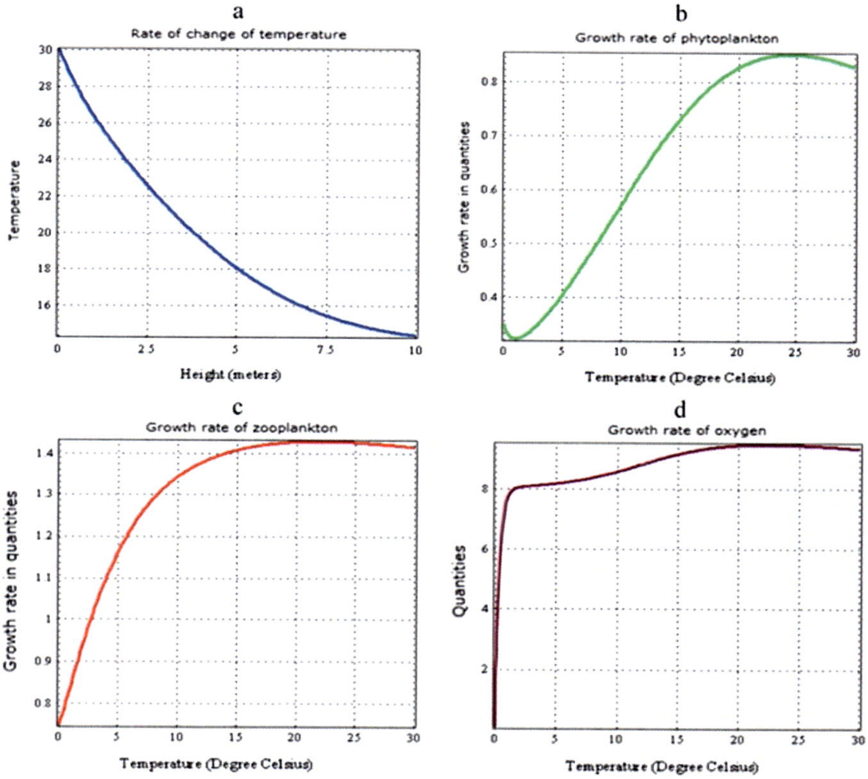

Fig. 4 Effect of temperature on planktonic ecosystem under $a = 30\,°\mathrm{C}$, $\kappa_1 = 0.83\,\mathrm{kl}^{-1}$, and $\kappa_2 = 0.320\,\mathrm{kl}^{-1}$

Figures 2, 3, and 4 represent that the ecosystem enriches gradually until the optimum temperature comes, and the system is optimum at the optimum temperature, and the system starts to decline for the high temperature (above optimal temperature).

4 Conclusions

In this study, a nonlinear mathematical model has been propounded and analyzed for the effect of temperature on the marine planktonic ecosystem. The model exhibits three equilibrium points where all the critical points will be stable under some conditions. We compute numerical simulation at the optimum temperature and a comparison has been shown in this section. The growth rate of phytoplankton at 25 °C is higher than any growth rate at any temperature. When the growth rate of phytoplankton increases, the growth rate of zooplankton rises, and consequently the production rate of oxygen arises. Thus, all the dynamical species reach to a stable relationship.

References

1. Temperature affecting the rate of photosynthesis. http://www.passmyexams.co.uk-/GCSE/biology/temperature-affecting-rate-of-photosynthesis.html. Last accessed 18 Jan 2019
2. Understanding the optimum temperature for plants. http://www.just4growers.com/stream/-temperaturehumidity-and-c02/understan-ding-the-optimum-temperature-for-plants.aspx. Last accessed 18 Jan 2019
3. Schabhuttl, S., Hingsamer, P., Weigelhofer, G., Hein, T., Weigert, A., Striebel, M.: Temperature and species richness effects in phytoplankton communities. Oecologia **171**(2), 527–536 (2012)
4. Striebel, M., Schabhuttl, S., Hodapp, D., Hingsamer, P., Hillebrand, H.: Phytoplankton responses to temperature increases are constrained by abiotic conditions and community composition. Oecologia **182**(3), 815–827 (2016)
5. Destania, Y., Jaharuddin, Sianturi, P.: Stability analysis of plankton ecosystem model affected by oxygen deficit. Appl. Math. Sci **9**(81), 4043–4052 (2015)
6. Khare, S., Kumar, S., Singh, C.: Modelling effect of the depleting dissolved oxygen on the existence of interacting planktonic population. Elixir Appl. Math **55**, 12739–12742 (2013)
7. Sekerci, Y., Petrovskii, S.: Mathematical modeling of plankton-oxygen dynamics under the climate change. Bull. Math. Biol. **77**(12), 2325–2353 (2015)
8. Promrak, J., Rattanakul, C.: Effect of increased global temperatures on biological control of green lacewings on the spread of mealybugs in a cassava field: a simulation study. Adv. Diff. Eq. **161**, 1–17 (2017)
9. Edwards, K.F., Thomas, M.K., Klausmeier, C.A., Litchman, E.: Phytoplankton growth and the interaction of light and temperature: a synthesis at the species and community level. Assoc. Limnol. Oceanogr. (ASLO) **61**, 1232–1244 (2016)
10. Sherman, E., Keith, J., Primeau, F., Tanouye, D.: Temperature influence on phytoplankton community growth rates. Glob. Biogeochem. Cycles 550–559 (2016)
11. Biswas, M.H.A., Rahman, T., Haque, N.: Modeling the potential impacts of global climate change in Bangladesh: an optimal control approach. J. Fundam. Appl. Sci. **8**(1), 1–19 (2016)
12. Akter, S., Islam, M.S., Biswas, M.H.A., Mandal, S.: A mathematical model applied to understand the dynamical behavior of predator prey model. Commun. Math. Model. Appl. **4**(3), 84–94 (2019)

SMART Asthma Alert Using IoT and Predicting Threshold Values Using Decision Tree Classifier

Anoop Kumar Prasad

Abstract Asthma is a chronic disease of the airways that transport air to and from the lungs. Yet, little full cure is available, but management methods can help a person with asthma lead a full and active life. Management of asthma before triggering can help in better treatment and long-term relief of this chronic disease. The proposed device collects the data and analyzes it as programmed. The device uses the concept of the Internet of things and data mining. Initially, the device is set up in the way that it can alert the patient to take quick-relief medication by sensing the configured trigger detectors and systemized in a prescribed way to alert about taking controllers. The critical threshold for environmental triggers asks the user to take medications before long exposure to that harmful environment which can result in asthma episodes. These way-controlled measures can be taken for asthma.

Keywords Asthma · Sensors · Data analyze · Controller · Asthma episode · Decision tree · Bolt IoT · Arduino

1 Introduction

Asthma is a common long-term inflammatory disease of the airways of the lungs. Effects include episodes of wheezing, coughing, chest tightness, and shortness of breath. These episodes may occur a few times a day or a few times per week. Asthma can be classified as mild, moderate, and severe asthma. Mild asthma patients have symptoms more than twice a week but not daily. Here, daily activities slightly get affected. Moderate asthma patients have symptoms daily. In this case, the daily activities get 50% affected. The use of medication becomes regular use for the patient once to twice daily. Severe asthma patients have an occurrence many times a day. The daily activities are 80% affected. The controller's usages become more than twice daily. Generally, asthma becomes uncontrolled if proper quick-relief medications are not taken when the environment triggers the patient. It is advised to visit doctor

---

A. K. Prasad (✉)
Computer Science and Engineering, Assam Science and Technology University, Royal School of Engineering and Technology, Guwahati, Assam, India
e-mail: anoopkprasad@rgi.edu.in

J. K. Mandal et al. (eds.), *Applications of Internet of Things*, Lecture Notes in Networks and Systems 137, https://doi.org/10.1007/978-981-15-6198-6_13

immediately. Foremostly, the device proposed here signifies the alert to use quick-relief medications by sensing the environmental triggers, and it is configured to sense using sensors. Secondly, the device has the capability to alert the reminder of taking controllers as prescribed to treatment patterns and quality of life (QOL) by using questionnaires designed for patients and physicians [1].

2 Background Study

The common symptom of asthma is wheezing—an audible piping that happens when air is inhaled, and it moves inside and outside of narrow airways. The narrowness of the bronchial tubes is a result of inflammation that causes the muscles surrounding the airways to tighten. Another common symptom of asthma is coughing. Nocturnal coughing is associated with asthma. The asthmatic patient is often report feeling of 'tight chested,' or lessen in breath. Asthma is an inveterate and often lifelong condition.

A need to confront defects in current asthma management are leading to a revaluation of the approach of personalized health care, which is strongly incentivized by the availability of new biologic treatments and methods for monitoring disease activity [2].

2.1 Earlier Developed Devices' Disadvantage

1. ADAM, the automated device for asthma monitoring is a device that quantifies symptoms in numbers, based on predetermined algorithms of symptom sounds including coughs and wheezes. It lacks in triggering prediction.
2. The 'spirometry' checks how well our lungs are working and performing by letting the patient blow out as hard, fast, and long in the apparatus and thus comparing it with the earlier observation of healthy people. Testing was taken again after medicine that opens the airways in our lungs if our results improve this is another sign of asthma.
3. Nitric oxide sensor helps in measuring the nitric oxide that is produced throughout the body, including the lungs, to fight inflammation and relax tight muscles. High levels of exhaled nitric oxide in the breath can mean that the airways are inflamed—one sign of asthma [3].

As result devices were not able to sense the environmental triggers.

2.2 The Causes of Asthma Exacerbations

Identifying triggers helps one to eliminate the exposure to them or, in instances, where that is not possible, counteract their effects [4]. Common environmental allergic triggers for asthma are:

- Cigarette smoke,
- Nitric oxide,
- Dust,
- Grass or pollen,
- Pet or cockroach dander, etc. [5].

The grass, pollen, pet, or cockroach dander can be outlined by maintaining a clean and hygienic environment. The cigarette smoke, nitric oxide, and dust are the factors that are mostly overseen by the patient which causes severe attacks of asthma. These attacks are slow and increase rapidly with the intake of these allergic triggers. When most people come across the heavy dust, the reflex action is simply to sneeze or cough. For the person with asthma, the response can be a full-blown asthma episode [4].

However, it is not possible for a patient using the inhaler to detect the triggers every time. To jog one's memory to clasp the inhaler (recommended by doctor) before the asthma attacks take place, our proposed model alerts for the timely intake of controllers and sensing the trigger and alerting the person to take quick-relief medications.

The word 'SMART' here denotes 'single maintenance and reliever therapy.' Asthma constitutes a clinical syndrome associated with the complex cause and subsequent development of an abnormal condition or of disease mechanisms that lead to a variable limitation of expiratory airflow and several clinical symptoms. These vary over time in their occurrence, frequency, and intensity. An asthma episode narrows and blocks the airways which make the patient uncomfortable to breathe and chest tightening.

Fast-acting medicines relieve constriction, whereas controller medicines work to prevent constriction from occurring in the first place by controlling inflammation. Controller medicines come in different forms, such as tablets, inhalers, and even injections. Controller medicines will not relieve symptoms during an asthma flare, so it is very important to understand the difference between the two: Controllers are long-term medications to prevent episodes, while quick-relief medications rapidly open the airways so the patient can breathe more comfortably once an episode has begun [4]. Untreated asthma limits the ability to live an active life, and still many asthmatics do not have the level of control over their asthma as they could have. In addition, about 50% of asthmatics use their inhaler. Most adults and children with asthma can obtain quite a good control of their disease using inhaled therapies. However, despite the optimization of standard therapies, patients with severe asthma, who can amount to 5–10% of the global population of asthmatic patients, may need an adjunctive biological treatment [6]

The proposed system ‘SMART Asthma Alert’ monitor the time inhaler is used to provide suitable medication to the patients once the reports are analyzed by the physicians. This involves air quality which is used to determine the quality of air in the patients’ environment. The impact of the environment in triggering the asthma attacks can be alerted to the patient before the condition of the patient gets severe. Asthma patient says that extreme temperatures are common nonallergic trigger. The regular healing with anti-inflammatory medication is necessary when they are not sensing symptoms.

3 Proposed Model

The proposed model here is to control episodes of asthma. With regular care, we can keep asthma under control as much as possible. To setup the device for the individual, we configure the installation (Fig. 1).

This process is followed by ‘data collection’ or prehistory records (if available) of the patient. Decision tree classifier is applied to the collected data (as patient adaptability differs from one-to-one. Accordingly, we set the prototype for the patient. The working block diagram is as follows (Fig. 2).

Fig. 1 Device setup for pilot testing

Fig. 2 Diagram representing the flow of the device. The concept that 'SMART Asthma Alert' uses are robust as shown

The model which the device follows is the 'prototype model.' This model gives us the flexibility of studying the progress and to change requirements as required (Fig. 3).

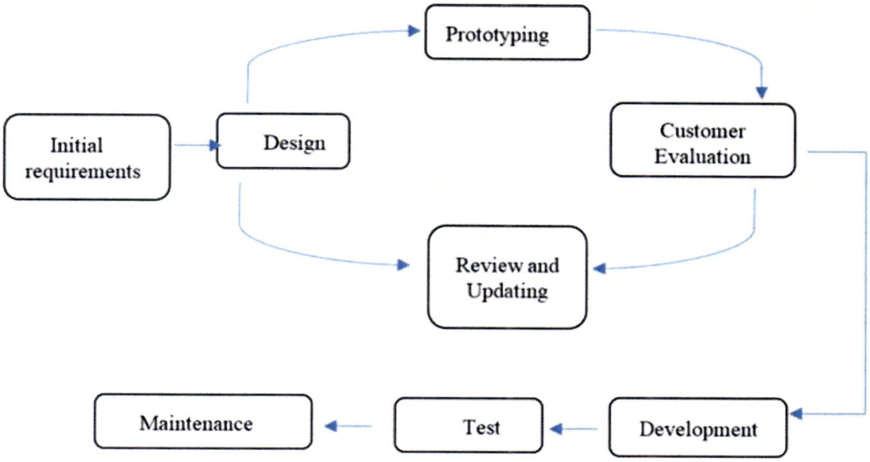

Fig. 3 Prototype model illustrating life cycle model which helps to produce a better quality of the product. It also helps in minimizing the chances of time and cost overrun

4 Design

Using ESP2866 model (Bolt IoT) or NodeMCU (A Lua-based firmware) or Raspberry Pi, we can connect GP2Y1010AU0F Optical Dust Sensor, 'NO' Gas Sensor and LM35 sensor. We interface Bolt IoT and Arduino. Using 'Bolt IoT module Arduino helper library,' we send analog signals from Arduino to Bolt. Using cloud computing, the data is processed online. We use Ubuntu server and set the criteria manually. Criteria are set by picking threshold values from looking into the prehistory records of the patient or finding it with the help of physician. The device uses the concept of the Internet of things for collecting the data. However, a mobile application is set up for the alert of taking quick-relief medications and notification alert for the reminder of controllers as prescribed. The later part uses the Internet for communication of machine to machine, i.e., the Bolt module model with any smartphone and configured mobile application to receive the push notification or email. This model is a low-cost Wi-Fi microchip with full TCP/IP stack and microcontroller capability.

Data mining concept of decision tree classifier helps us understand the nature of the triggers and thus helps us in setting up the threshold for the patient. Decision trees are an important tool for developing classification or predictive analytics models related to analyzing big data or data science.

The presence of redundant attributes does not adversely affect the construction of the decision tree. Here, while collecting data, we get many redundant sets of data points too, e.g., missing values by the sensors and noise values. The prediction of another episode of asthma attacks can be further analyzed using machine learning prediction.

Initially, the device is set up in a way that it can alert the patient to take quick-relief medication by sensing the configured trigger detectors and systemized in a prescribed way to alert about taking controllers. The critical threshold for environmental triggers asks the user to take medications before long exposure to that harmful environment which can result in asthma episodes. The model can be built as a portable one to be handy for the patient to carry it attaching it with the backpack, school bag, or car seat. At the same time, configuring it in such a way that the device can sense the environment easily. The device can be provided energy using a 3.7 V battery 2000mAh, which helps in the smooth functioning of the device for more than 24 h. The device serves the purpose of the vision to improve global health by enabling the correlation between personal health and personal environment.

5 Tests and Results

Pilot testing of the project helps in understanding the relation between environmental triggers and the device.

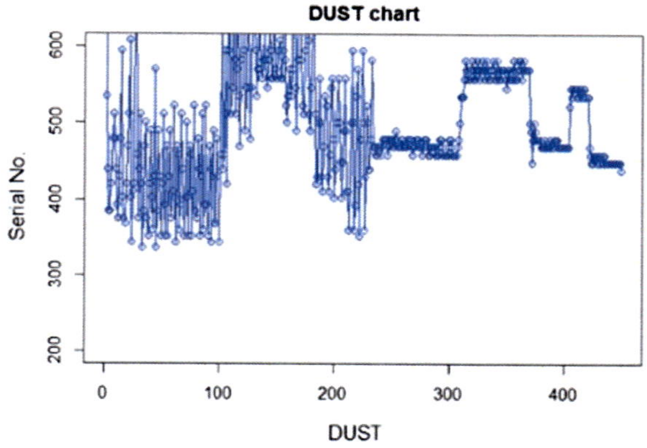

Fig. 4 'Sample DUST dataset' representing the presence of dust

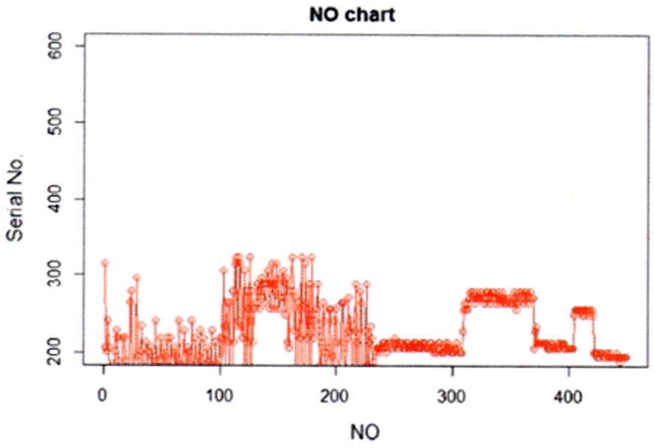

Fig. 5 'Sample NO dataset' representing the presence of 'NO'

5.1 *Dataset*

The dataset is a sample dataset used here. It is an exemplary form. The data values vary from that of real results of the individual allergic triggering data points. The datasets are preprocessed by labeling them into two categories, harmful, and unharmful. The label is termed as—'HAMRFUL' or 'NOHARM' which indicates the binary form of 1 or 0 to help in classification of the dataset and understanding the threshold (Figs. 4, 5, 6, and 7).

5.2 *Analyses*

In this subsection, the model initially does not understand the patient requirements or the underlying technical aspects. However, these can be identified before the projects

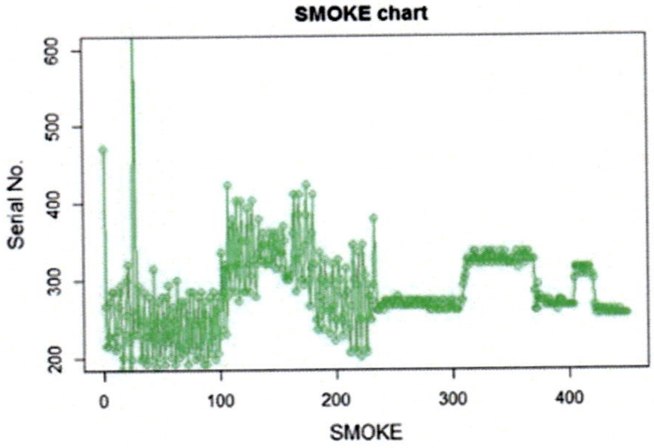

Fig. 6 'Sample SMOKE dataset' representing the presence of smoke

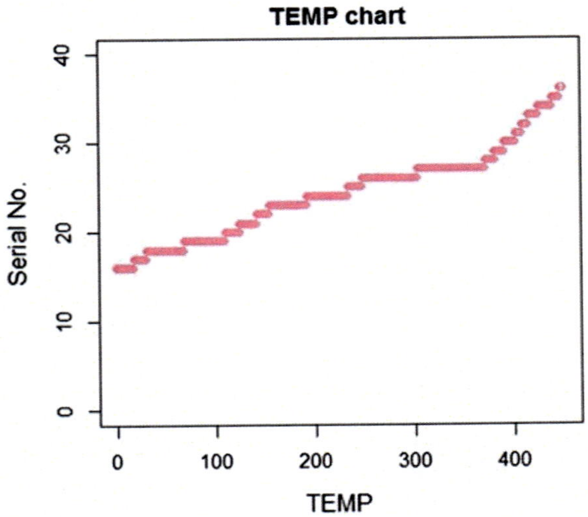

Fig. 7 'Sample TEMP dataset' representing the presence of temperature

start. By knowing the patient's allergic trigger threshold, we set with datum whether it is harmful or not for the patient. This process of noting down of each datum is patient specific. The sample dataset shows the presence of DUST, NO, SMOKE, and TEMPERATURE, respectively. The dataset is then complete with more than 400 data points. The dataset needs to be taken in the metric system. Here, the unit used is 'PPM' for dust, nitric oxide, and smoke. For temperature data points unit used is 'Celsius.' The decision tree classifier is applied to the dataset having the above-mentioned factors to get the threshold value for the individual patient.

Decision tree learning is one of the predictive modeling approaches used in statistics, data mining, and machine learning. It uses a decision tree to go from observations about an item to conclusions about the item's target value (Fig. 8).

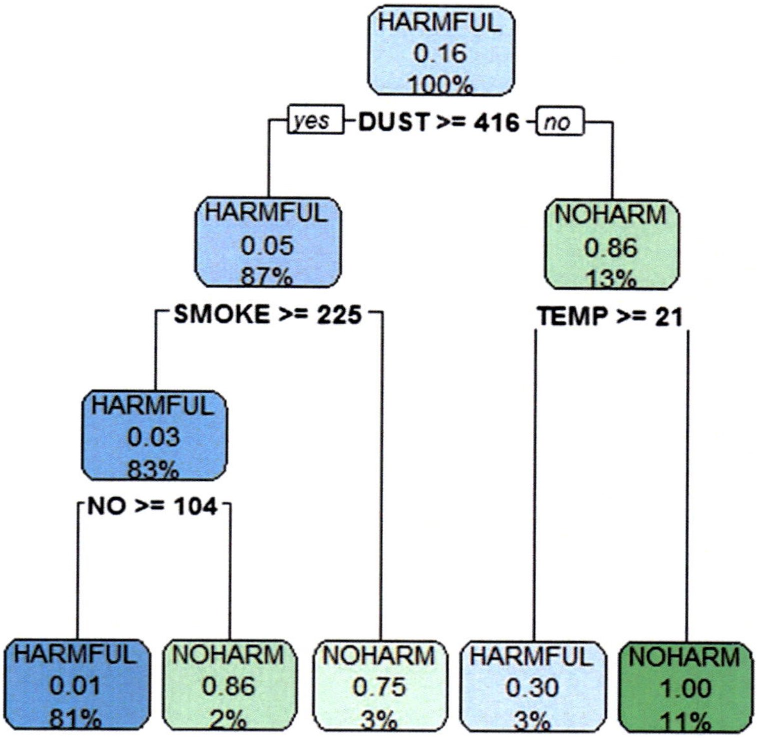

Fig. 8 Decision tree classifier is used, and it has produced the classified result. The nature of the factors and its detection for individual helped us knowing the effects of each factor as 'harmful' and 'not harmful.' The accuracy obtained is 97.7%

6 Conclusion

The word 'SMART' here denotes 'single maintenance and reliever therapy.' Controllers are long-term medications to prevent episodes of asthma, while quick-relief medications immediately open the narrow airways to help the patient breathe more comfortably once the episode has begun. The device sends an email alert and pushes a notification to the user's account or device. Thus, the device is helpful worldwide to prevent asthma patient to delay in their daily activities.

Future Scope

We understand the future scope of the device and how it can be more advanced with the trending technologies.

Using nanotechnology, we can create specific designs for the product which will be tiny to get fit with wearable devices. Research work for supplying the energy to the device can be generated using solar panels or harvest power from the human body which relies on heat and motion, i.e., conversion of potential energy to electrical energy. 'SSP working model solar panel with 3.8 V' can be used as energy supplier for the product. This will help in cutting the cost of the device and long-term durability. This write-up gives enhancements to advance environmental health research and policy and strengthen clinical trends.

Acknowledgements I appreciate Ritika Jain and Mahammad Rowshan Chowdhury for their involvement in the questionnaire. I gratefully acknowledge my parents the teachers of Royal Global University, Computer Science and Engineering Department to guide me from the initial days to teach me how to write a paper and learn to present it. My heartfelt thanks to Ms. Gitimoni Talukdar, Ms. Ishita Chakrabarty, Ms. Ankita Goyal Agarwal, Mr. Debashish Mishra, Mr. Nayan Jyoti Kalita, and Mr. Sasank Boruah to show me the ways to write the paper and inspiring a hope to advance myself in this field. My sincere thanks to Mr. Saurabh Sutradhar and Mr. Manoj Kumar Sarma to guide me to purposefully write this paper.

I thank Dr. Aniruddha Deka, HoD of Computer Science Department for encouraging me to publish it. I will always be thankful to my parents and sister for constant support.

References

1. Ohta, K, Tanakam, H., Tohda, Y., et.al: Asthma exacerbations in patients with asthma and rhinitis: factors associate with asthma exacerbation and its effect on QOL in patients with asthma and rhinitis. doi: https://doi.org/10.1016/j.alit.2019.04.008
2. Thomas, M.: Why aren't we doing better in asthma: time for personalised medicine? NPJ. Prim. Care Respir. Med. **25**, 15004 (2015)
3. Clinic, M.: Nitric oxide test for asthma
4. Cook, G.W.: So, your doctor says you have asthma. Asthma Mag. **10**(2), 18–20 (2005). https://doi.org/10.1016/j.asthmamag.2005.02.003
5. Huan, J., Pansare, M.: New treatments for asthma
6. Pelaia, C., Calabrese, C., Terracciano, R., de Blasio, F., Vatrella, A., Pelaia Ther, G.: Omalizumab, the first available antibody for biological treatment of severe asthma: more than a decade of real-life effectiveness. Adv Respir Dis 12:1–16 (2018). https://doi.org/10.1177/1753466618810192

Object-Oriented Modeling of Cloud Healthcare System Through Connected Environment

Subhasish Mohapatra, Komal Paul, and Abhishek Roy

Abstract Advancement of information and communication technology (ICT) has facilitated electronic communication among its users having dispersed geographical location. This concept of electronic communication may be implemented in multivariate service sectors to cater services to end user, i.e., Citizen. India being a developing nation have to afford huge establishments and manpower to deliver services regularly to its Citizen. Due to this situation, enormous amount of recurring expenditure is mounted over the financial structure of nation. Moreover, the services delivered through conventional mode takes sufficient time to reach the intended user, particularly, in case of distantly located user. To solve these issues, electronic mode of message communication may be adopted to provide electronic services in timely and budget-friendly manner. The only concern of this approach is security of sensitive information which is transmitted through public communication channel, i.e., Internet. To resolve this issue, hybrid cryptographic security protocols should be used to ensure privacy, integrity, non-repudiation and authentication (PINA) during its implementation in real-world scenario. Furthermore, to provide an user friendly system, a Citizen-centric single window interface have been modeled. As the primary objective of this paper, authors will extend it further to deliver cloud healthcare facilities to Citizen through the proposed single window interface, i.e., multipurpose electronic card (MEC). To simulate the real-world implementation, in this paper, authors have performed object-oriented modeling (OOM) of proposed cloud healthcare system.

S. Mohapatra · K. Paul · A. Roy (✉)
Department of Computer Science & Engineering, Adamas University, Kolkata, India
e-mail: dr.aroy@yahoo.com

S. Mohapatra
e-mail: mohapatra.subhasish@gmail.com

K. Paul
e-mail: komal.paul88@gmail.com
URL: http://adamasuniversity.ac.in/

A. Roy
International Association of Engineers, Hong Kong, China

Cryptology Research Society of India, ISI Kolkata, Kolkata, India

J. K. Mandal et al. (eds.), *Applications of Internet of Things*, Lecture Notes in Networks and Systems 137, https://doi.org/10.1007/978-981-15-6198-6_14

Keywords Cloud computing · Cloud healthcare · Object-oriented modeling (OOM)

1 Introduction

Advancement of information and communication technology (ICT) has facilitated electronic communication among its users having dispersed geographical location. Internet as the communication medium helps to transmit electronic messages among its users connected to each other. This concept of electronic communication may be implemented in multivariate service sectors to cater services to end users, i.e., Citizen. India being a developing nation have to afford huge establishments and manpower to deliver services regularly to its Citizen. Due to this situation, enormous amount of recurring expenditure is mounted over the financial structure of nation. Moreover, the services delivered through conventional mode takes sufficient time to reach the intended user, particularly, in case of distantly located user. To solve these issues, electronic mode of message communication may be adopted to provide electronic services in timely and budget-friendly manner. The only concern of this approach is security of sensitive information which is transmitted through public communication channel, i.e., Internet. To resolve this issue, hybrid cryptographic security protocols should be used to ensure privacy, integrity, non-repudiation and authentication (PINA) during its implementation in real-world scenario. In this paper, a Citizen-centric single window platform have been modeled using cloud computing [1–4], to deliver multivariate electronic services [5] to end users. To expand it further, in this paper, authors have proposed cloud healthcare [6–10] system for delivery of healthcare facilities to Citizen (i.e., patient) using an electronic instrument (i.e., multipurpose electronic card).

The rest of the paper is organized as mentioned below:

1. Phase I: As a part of Cloud Governance [2] transaction, Citizen communicates with Government to avail medical facilities through Citizen to Government (C2G) type of transaction as shown in Fig. 1. Government verifies Citizen to avail desired facility, which is briefly discussed in Sect. 2.
2. Phase II: As a part of proposed cloud healthcare system, Citizen will communicate with healthcare services to avail desired medical facilities. In this paper, authors have focused in this phase of transaction only. Furthermore, to simulate the real-world implementation, the object-oriented modeling (OOM) of proposed cloud healthcare system is shown through various figures like, Figs. 2, 3, 4, 5, 6, 7, 8, 9, 10, 11 and 12, which are explained in Sects. 3 and 4, respectively.
3. Phase III: Sect. 5 finally draws the conclusion of this work along with its future scope, like payment of medical expenses, etc.

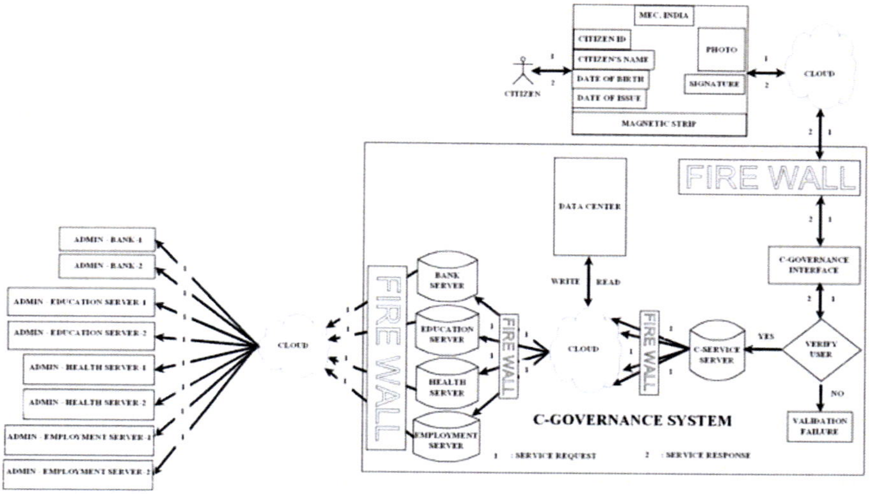

Fig. 1 Schematic diagram of cloud governance system

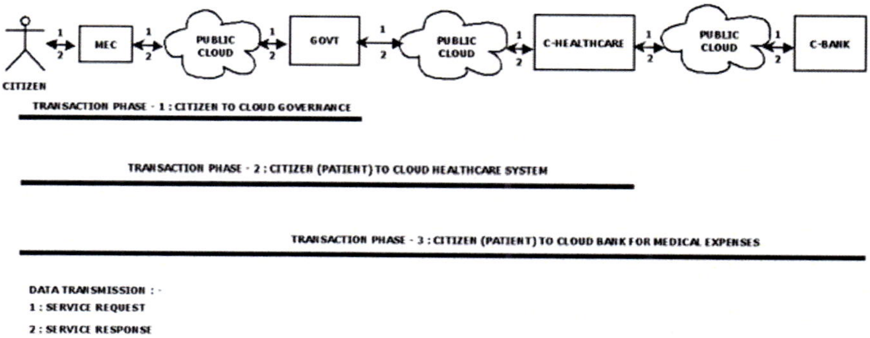

Fig. 2 Conceptual diagram of cloud healthcare system

2 Origin of Work

Figure 1 shows the single window-based Citizen-centric cloud governance system [2, 11], which contains following three primary participants:

1. Citizen: Citizen uses multipurpose electronic card (MEC) to avail electronic services under the jurisdiction of Government.
2. Government: Government monitors the SERVICE REQUEST of Citizen and corresponding SERVICE RESPONSE of service providers under its jurisdiction.
3. Service Provider: Service provider represents any entity or organization which provide desired SERVICE RESPONSE to the Citizen. The servers are shown in

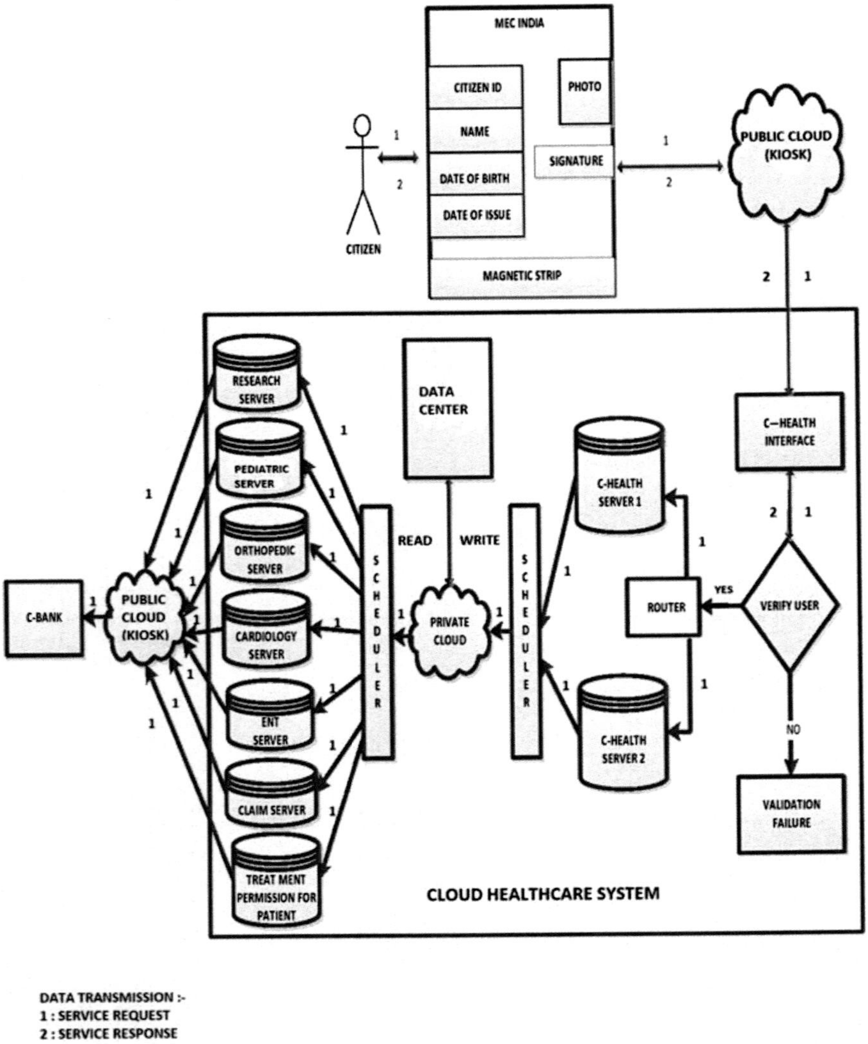

Fig. 3 Schematic diagram of cloud healthcare system

Fig. 1 denote those multifaceted services currently available for Citizen, which can also be extended further to increase the number of facilities.

As these primary actors communicate among themselves through public cloud, there is wide window available for improvement of data security.

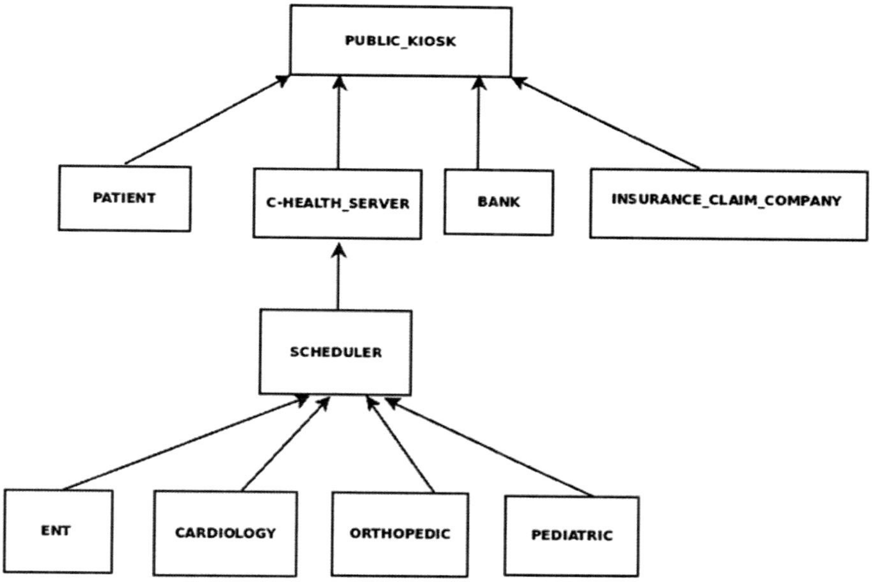

Fig. 4 Block diagram of cloud healthcare system

Fig. 5 Class diagram of patient

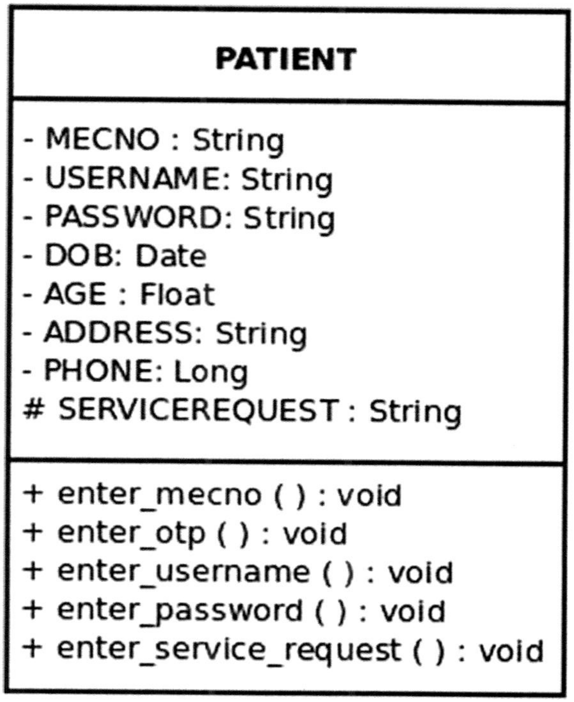

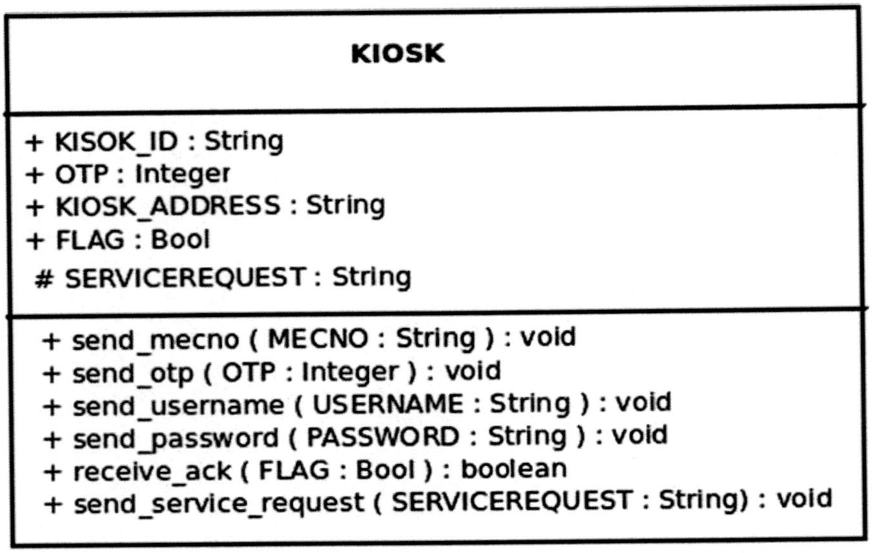

Fig. 6 Class diagram of public kiosk

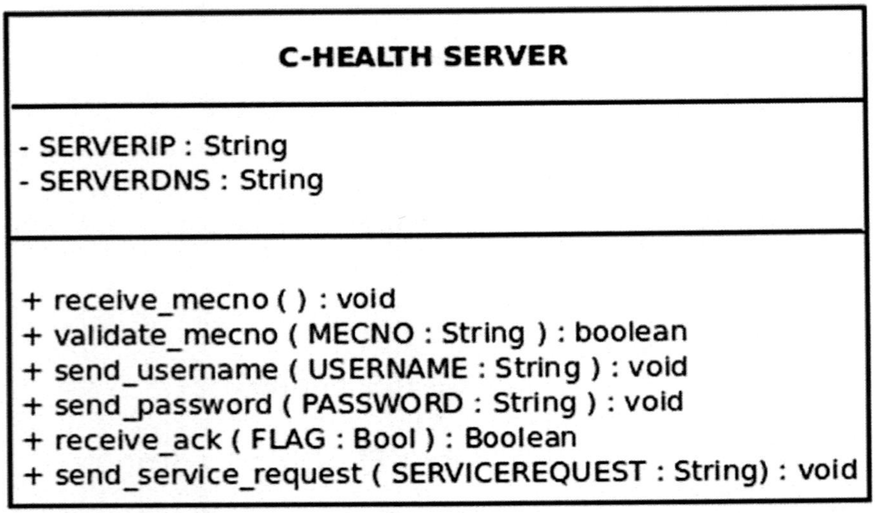

Fig. 7 Class diagram of cloud health server

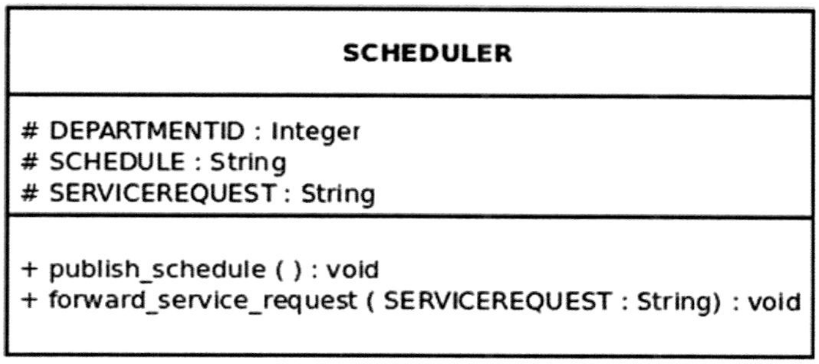

Fig. 8 Class diagram of scheduler

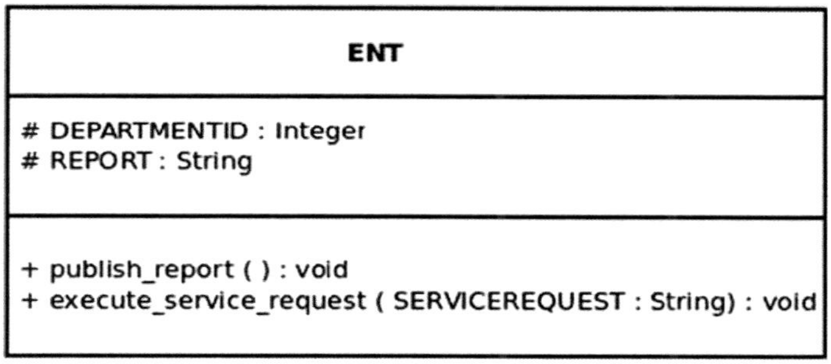

Fig. 9 Class diagram of ENT department

2.1 Cloud Governance System

The electronic transaction is shown in Fig. 1 during Citizen to Government (C2G) type of transaction is stated below:

1. Citizen initiates transaction with Government using multipurpose electronic card (MEC). Multipurpose electronic card (MEC) denotes a single window interface to access electronic services, which also helps Government to identify Citizen using an unique identification number.
2. Citizen transmit unique parameters and SERVICE REQUEST to Government through Path-1 of Fig. 1.
3. Government verifies the identity of Citizen.
   (a) In case of successful verification, transaction proceeds toward Step-4.
   (b) In case of unsuccessful verification, the transaction is aborted and Citizen is informed through system timeout via Path-2 of Fig. 1.
4. SERVICE REQUEST of Citizen is analyzed in cloud service server to understand the exact service requested by Citizen.

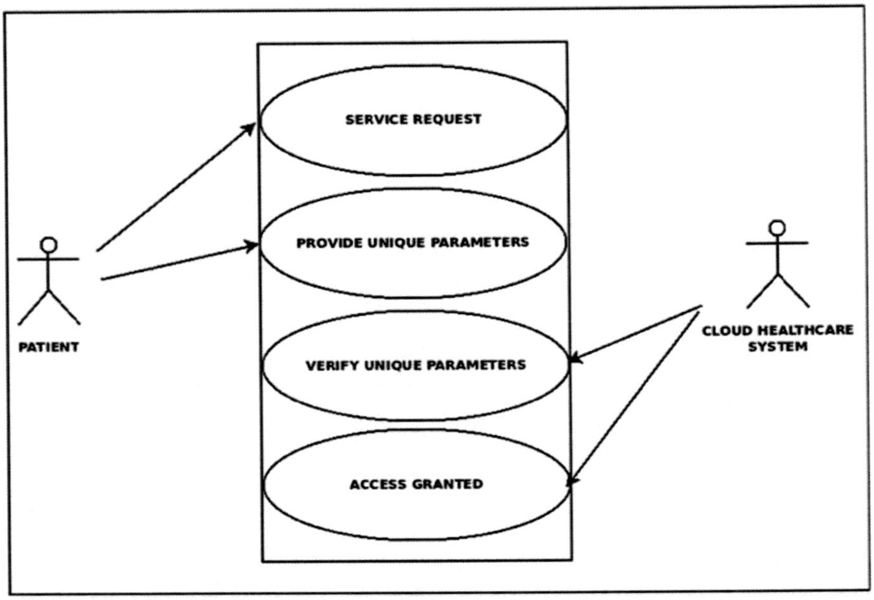

Fig. 10 Use case diagram of patient and cloud healthcare system

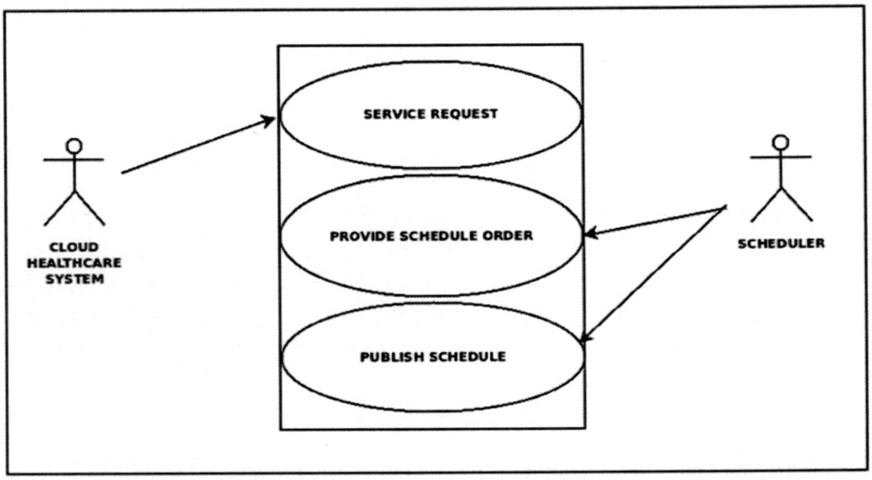

Fig. 11 Use case diagram of cloud healthcare system and scheduler

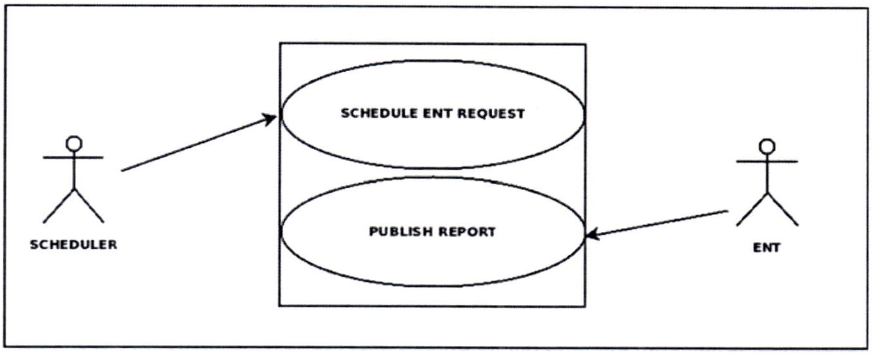

Fig. 12 Use case diagram of scheduler and ENT department

5. Data center attached to the private cloud of cloud governance system performs corresponding READ and WRITE operations for SERVICE REQUEST of Citizen.
6. Service servers like bank server, education server, health server, employment server receives the SERVICE REQUEST of Citizen through private cloud of cloud governance (i.e., C-Governance) system and route it to the exact third-party service provider through public cloud. Steps explained through Step-1 to Step-6 are represented by Transaction Phase-1 of Fig. 2.
7. Service provider receives SERVICE REQUEST of Citizen through public cloud and implement it using its internal mechanism. This step is represented by Transaction Phase-2 of Fig. 2.
   Transaction Phase-3 of Fig. 2 denotes the expenses incurred by Citizen for availing electronic facilities, which is the future scope of this work.

As a comparison to this concept discussed in Sect. 2, we have extended it further by introducing scheduler within our proposed cloud healthcare system, which is discussed in Sects. 3 and 4, respectively. This scheduler will help to maintain a proper queue of **SERVICE REQUEST** of patient (i.e., Citizen) and generate corresponding **SERVICE RESPONSE** accordingly. Furthermore, to provide an optimal balance between aforementioned concept, authors come up with detailed class diagram of each module involved in this work, each UML class of healthcare system [12] provide an empirical evidence of attributes that act as building block of model. Hence, UML-based object-oriented (OO) analysis of model will definitely improve design quality of software prior to its implementation.

3 Proposed Cloud Healthcare System

India is striving hard to provide world class healthcare facilities to its populace within an affordable budget. The situation is highly critical for remote locations where basic healthcare amenities like availability of efficient doctors, medicine, vaccines, trained nursing staff, etc., inadequate in nature. Due to this reason, patient with critical condition get expired while physically traveling long distance to reach the hospital, pregnant women faces severe health issues while delivering baby, thereby leading to untimely death of baby or mother or both in worst cases. In this condition, it is really a challenge to deliver advanced medical facilities and consultation to patient (i.e., Citizen) in timely and cost-effective manner. To meet up this gap between SERVICE REQUEST of Citizen and corresponding SERVICE RESPONSE, we have opted for technology based solution and proposed a cloud healthcare system, which is shown in Figs. 2 and 3, respectively. Among the multiple phases shown in Fig. 2, Transaction Phase-2 is specifically elaborated in Fig. 3, which is implemented only after getting necessary clearance from Government through Citizen to Government (C2G) type of electronic transaction. Critics may raise question about involvement of Government for availing medical facilities. Since Government is accountable for maintenance of health and hygiene of populace, its involvement during our proposed cloud healthcare system will provide additional advantage to patient (i.e., Citizen) and the society as a whole. In Fig. 3, we have shown Citizen (i.e., patient) to cloud healthcare (i.e., C-Healthcare) service (C2H) type of transaction, which is described below:

1. Citizen (i.e., patient) side:
   (a) Citizen initiates cloud healthcare transaction using public cloud (Kiosk) through Path-1 of Fig. 3.
      i. Citizen provides unique parameter using multipurpose electronic card (MEC), through which cloud healthcare service provider identifies the patient.
      ii. Citizen provides SERVICE REQUEST to avail specific healthcare facility.
2. Cloud healthcare (i.e., C-Healthcare) service provider side:
   (a) C-Healthcare service provider receives information of Citizen (i.e., patient) stated in Step-1(a)i and Step-1(a)ii using Path-1 of Fig. 3.
   (b) C-Healthcare service provider verifies the identity of Citizen (i.e., patient).
      i. In case of unsuccessful verification, the SERVICE REQUEST of Citizen (i.e., patient) is aborted and intimated through system timeout using Path-2 of Fig. 3.
      ii. In case of successful verification, the C-Healthcare transaction proceeds further through Step-2c.
   (c) Router which helps to balance huge data transmission load over the network, receives SERVICE REQUEST of patient and en route it to C-Healthcare

server. In Fig. 3, multiple C-Health servers are shown to demonstrate load balancing for huge amount of SERVICE REQUEST send by Citizen (i.e., patient), which can explored further by application of distributed databases during this phase of electronic transaction.

(d) The scheduler connected with all the C-Healthcare servers handles the SERVICE REQUEST to generate a SERVICE QUEUE on FIRST COME FIRST SERVE basis for sequential execution of SERVICE REQUEST.

(e) Private Cloud of proposed C-Healthcare System, which is attached with the scheduler receives SERVICE REQUEST of Citizen (i.e., patient) and performs necessary READ and WRITE operation over the data center.

(f) Another scheduler attached with the private cloud transmits the SERVICE REQUEST to respective C-Healthcare server through Path-1 of Fig. 3.

(g) Specific C-Healthcare servers like research server, pediatric server, orthopedic server, cardiology server, ENT server, claim server, treatment permission for patient server, etc., executes the SERVICE REQUEST of patient (i.e., Citizen) to deliver the desired C-Healthcare facility. Apart from research server, all other servers will directly engage with Citizen (i.e., patient) for delivery of medical facilities, whereas research server will study these transactions for further enhancement of proposed cloud healthcare system.

(h) The medical expenses incurred by Citizen (i.e., patient) in this cloud-based service delivery model is considered as future scope of work.

As our proposed cloud healthcare system have to perform under real-world scenario, its robustness and dynamic feature should be measured properly before investing hard earned money. Hence, its dynamic features are explained using object-oriented modeling (OOM) in Sect. 4 of this paper.

4 Object-Oriented Modeling (OOM) of Cloud Healthcare System

The primary diagrams used to perform object-oriented modeling (OOM) of proposed cloud healthcare system are explained below:

1. Figure 4 of Sect. 4.1 shows the basic structure of the proposed C-Healthcare system.
2. Figures 5, 6, 7, 8 and 9 of Sect. 4.2 show the static structure of the proposed cloud healthcare system using class diagram.
3. Figures 10, 11 and 12 of Sect. 4.3 discuss the primary actors of the proposed system using use case diagram.

These initial drafts of proposed cloud healthcare system will be enhanced to include further healthcare facilities for its end user, i.e., patient (i.e., Citizen).

4.1 Block Diagram

The block diagram of proposed cloud healthcare system shown through Fig. 4 is explained below:

1. PUBLIC_KIOSK: It represents the public cloud, which facilitates electronic communication between PATIENT (i.e., Citizen), C-HEALTH_SERVER, BANK and INSURANCE_CLAIM_COMPANY.
2. PATIENT: It represents the Citizen who sends specific SERVICE REQUEST to avail electronic healthcare facility.
3. C-HEALTH_SERVER: It represents the servers of proposed Cloud Healthcare system which implements the SERVICE REQUEST of Citizen (i.e., patient). It further categorized into the following blocks:
   (a) SCHEDULER: It represents the internal component of proposed Cloud Healthcare System, which publish collision-free schedule of patientś SERVICE REQUEST (i.e., patient) and forward it to service server of respective medical unit (i.e., department) like ENT, CARDIOLOGY, ORTHOPEDIC, PEDIATRIC, etc.
      i. ENT: It represents those medical services available for Citizen (i.e., patient) related to ear, nose and throat (ENT).
      ii. CARDIOLOGY: It represents those medical services available for Citizen (i.e., patient) related to cardiovascular system of human being.
      iii. ORTHOPEDIC: It represents those medical services available for Citizen (i.e., patient) related to deformities of bones and muscles.
      iv. PEDIATRIC: It represents those medical services mainly available for children.

      The medical facilities discussed above may be expanded further depending on the SERVICE REQUEST of Citizen (i.e., patient), which is considered as future scope of this work.
4. BANK: It represents the third-party entity (i.e., bank) using which Citizen (i.e., patient) make payment of all medical expenses.
5. INSURANCE_CLAIM_COMPANY: It represents the third-party entity (i.e., insurance company) using which Citizen (i.e., patient) make payment of medical expense, in case the patient is under any health insurance coverage.
   The involvement of BANK and INSURANCE_CLAIM_COMPANY within our proposed cloud healthcare system will be explored in the future.

4.2 Class Diagram

Figure 5 shows the essential parameters of patient using its class diagram. Figure 6 shows the essential parameters of public kiosk using its class diagram. Figure 7 shows the essential parameters of proposed cloud healthcare server using its class diagram. Figure 8 shows the essential parameters of Scheduler using its Class Diagram. Figure 9 shows the essential parameters of ENT department using its class diagram. Other medical units (i.e., departments) as shown in Fig. 4 will also have their similar class diagrams. Though Bank and Insurance Claim company are shown in Fig. 4 for complete visualization of the proposed cloud healthcare system, it will be considered as future scope of this work.

The use case diagrams of proposed cloud healthcare system are shown in Sect. 4.3.

4.3 Use Case Diagram

This section shows the interaction between the primary actors of proposed cloud healthcare system in a sequential manner. Figure 10 shows the interaction between patient (i.e., Citizen) and cloud healthcare system using its use case diagram. Figure 11 shows the interaction between proposed cloud healthcare system and its internal service scheduler for generation of service schedule mainly to avoid the deadlock situation of multiple SERVICE REQUEST. Figure 12 shows the interaction between internal service scheduler and specific healthcare unit (i.e., department like ENT, etc.,) for final execution of the SERVICE REQUEST. Other medical units (i.e., departments) as shown in Fig. 4 will also have their similar use case diagrams.

5 Conclusion

Authors sincerely admit that explanation through class diagram and use case diagram is insufficient for object-oriented modeling (OOM) of any electronic service delivery model. However, within the limited scope, in this paper, authors have proposed a user-friendly multivariate electronic healthcare service delivery model and explained its basic structure. Further explanation using metrics for object-oriented design (MOOD), incorporation of additional healthcare facilities and subsequent payment of medical expenses through electronic banking transaction (as shown through Transaction Phase-3 of Fig. 2) may be considered as future scope of this work. To conclude, this paper has addressed **SERVICE REQUEST** of patient, whereas the explanation of **SERVICE RESPONSE** in broader perspective will be the main objective of next work.

References

1. Biswas, S., Roy, A.: An intrusion detection system based secured electronic service delivery model. In: 3rd International Conference on Electronics Communication and Aerospcace Technology (ICECA 2019), pp. 1712–1717. IEEE Conference Record # 45616, ISBN 978-1-7281-0167-5, India (2019)
2. Roy, A.: Smart delivery of multifaceted services through connected governance model. In: 3rd International Conference on Computing Methodologies and Communication (ICCMC 2019), pp 493–499. IEEE Conference Record # 44992 ISBN 978-1-5386-7807-7, India (2019)
3. Singh, M., Srivastava, V.M.: Multiple regression based cloud adoption factors for online firms. In: 2018 International Conference on Advances in Computing and Communication Engineering (ICACCE 2018), pp. 147–152. IEEE, New York. https://doi.org/10.1109/icacce.2018.8457722 ISBN 978-1-5386-4485-0/18 (2018) Paris
4. Harfoushi, O., Akhorshaideh, A.H., Aqqad, N., Janini, M.A., Obiedat, R.: Factors affecting the intention of adopting cloud computing in Jordanian hospitals. Commun. Network **8**(2), 88–101 (2016). https://doi.org/10.4236/cn.2016.82010
5. Khatun, R., Bandopadhyay, T., Roy, A.: Data modelling for e-voting system using smart card based e-governance system. Int. J. Inf. Eng. Electron. Bus. **9**, 45–52 (2017). https://doi.org/10.5815/ijieeb.2017.02.06
6. Abouelmehdi, K., Beni-Hessane, A., Khaloufi, H.: Big healthcare data: preserving security and privacy. J. Big Data **5**(1), 1–18 (2018). https://doi.org/10.1186/s40537-017-0110-7
7. Mahalakshmi, M.V., Shrivakshan, G.T.: An efficient cloud computing security in healthcare management system. Int. J. Adv. Res. Comput. Sci. Software Eng. **7**(8), 185–192 (2017)
8. Hanen, J., Kechaou, Z., Ayed, M.B.: An enhanced healthcare system in mobile cloud computing environment. Vietnam J. Comput. Sci. **3**(4), 267–277 (2016)
9. Lee, T.: Mobile healthcare computing in the cloud. In: Mobile Networks and Cloud Computing Convergence for Progressive Services and Applications, pp. 275–294. IGI GLOBAL (2014). https://doi.org/10.4018/978-1-4666-4781-7.ch015 ISBN13: 9781466647817
10. Zhang, R., Liu, L:. Security models and requirements for healthcare application clouds. In: 2010 IEEE 3rd International Conference on Cloud Computing, pp. 268–275. https://doi.org/10.1109/CLOUD.2010.62 ISBN 978-0-7695-4130-3/10
11. Roy, A.: Object-oriented modeling of multifaceted service delivery system using connected governance. In: Jena A., Das H., Mohapatra D. (eds) Automated Software Testing. ICDCIT 2019. Services and Business Process Reengineering, pp. 1–25. Springer, Singapore. (2020). https://doi.org/10.1007/978-981-15-2455-4_1
12. Singh, I., Kumar, D., Khatri, S.K..: Improving the efficiency of e-healthcare system based on cloud. In: 2019 Amity International Conference on Artificial Intelligence (AICAI), pp 930–933. IEEE, Dubai, United Arab Emirates (2019)

Estimating RNA Secondary Structure by Maximizing Stacking Regions

Piyali Sen, Debapriya Tula, Suvendra Kumar Ray, and Siddhartha Sankar Satapathy

Abstract Various rudimentary cellular functions that are carried out in an organism are dependent on RNA secondary structure. Thus, accurate prediction of RNA secondary structure is becoming an increasing interest. There are several methods in the literature that predict the secondary structure. In this paper, a maximum independent set (MIS) approach to predict the RNA secondary structure is presented. We find all possible secondary structure which has maximum base pairs using MIS on circle graph. We not only concentrate on maximizing base pairs but maximizing stacking regions, as it reinforces the stability of secondary structure. We also compare the suboptimal structures using stacking energy and then Tinocos stability number. The output of our algorithm could be more than one secondary structure, as in real-life scenario, the secondary structures may have different sets of base pairs with similar energy level. We also have provided a Web portal named TU Web server available at http://14.139.219.242:8003/rna_struct to visualize predicted RNA secondary structure.

P. Sen · S. S. Satapathy (✉)
Department of Computer Science and Engineering, Tezpur University, Tezpur, Assam 784028, India
e-mail: ssankar@tezu.ernet.in

P. Sen
e-mail: piyalisen18@gmail.com

D. Tula
Department of Computer Science and Engineering, IIIT, Sri City, Chittoor, Andhra Pradesh, India
e-mail: tula.deb011@gmail.com

S. K. Ray
Department of Molecular Biology and Biotechnology, Tezpur University, Tezpur, Assam 784028, India
e-mail: suven@tezu.ernet.in

J. K. Mandal et al. (eds.), *Applications of Internet of Things*, Lecture Notes in Networks and Systems 137, https://doi.org/10.1007/978-981-15-6198-6_15

Keywords Secondary structure of RNA · Maximum independent set · Circle graph

1 Introduction

The primary structure of RNA is a nucleotide sequence which is single stranded in nature, where the nucleotides are of four types namely A, U, G, and C. Primary structure of RNA may not exist in a stable condition on its own, so the nucleotides have a tendency to pair among themselves to form base pair, where the possible pairs are G:C, A:U, and G:U, thus they fold to form a secondary structure. In case of tRNA and rRNA, the secondary and tertiary structures play important role for their functions, whereas in case of mRNA, the primary structure is important for the function. There are examples of RNA secondary secondary structure playing important role in replication control in single-stranded RNA virus [1], and mRNA structure motifs are known to have regulatory role on gene expression [2], binding of drug molecules to the structure of viral RNA, and translational control in RNA [3]. Therefore, estimating the accurate structure of RNA is thus interesting.

There are quite a few ways to determine RNA secondary structure that exist in literature which are briefly described in this section. NMR and X-ray crystallography being laboratory experiments lack in speed, are difficult and expensive to run different samples of RNA time and time again. Hence, it entices toward computational simulation to estimate RNA structure that is close to real. There are broadly two ways to determine RNA secondary structure. First, using multiple homologous strains of RNA or similar RNA sequences [4–7]. This method is reported as one of the widely accepted methods. But the shortcoming can be inadequacy of multiple strains for RNA. Second approach is by using only single strain, most notably using dynamic programming approach which is based on scoring system, and using free energy minimization [8–10], stochastic context-free grammar approach which is based on probability of base pairs [11], genetic algorithm selects the structure following a stepwise procedure and chooses the most fit structure [12], backtracking of path matrix [13], and thermodynamic RNA prediction [14].

Another approach is by finding the near-maximum independent set (MIS) of chords of a circle graph, where the nucleotides are placed on the circumference of circle graph. Base pairs are represented as the chords in the circle graph. MIS gives the largest number of vertices that are not adjacent to each other. In real scenario, one base pairs with exactly one base if any, and there would be no intersection of base pairs. So, a planar circle graph with maximum number of chords is supposed to provide suitable RNA secondary structure. To determine MIS of a graph is known to be NP-complete [15]. Still, there exist some methods that determine MIS [16, 17]. Parallel approach to determine MIS has also been suggested in literature [1, 3, 18]. This method is based on a single neuron model, which iterates over few hundred iterations to find the MIS. But some of the limitations of this method are, some parameters needs to be set at the start. On every run, these parameters need to be changed, that would give new MIS, which further needs to be compared with previous runs and to keep the optimal one. If the number of bases in RNA sequence is high enough, then one single run takes large amount of time. Selection of parameters is also a concern, as the results do not follow any definite pattern, so on what interval should we increase or decrease the parameters are a question.

In our approach to find MIS, we used igraph Python package to identify all possible MIS on a single run. The algorithm used is explained in method section to choose secondary structure having large proportion of sequence coming under stems. Because of limitation of computing power of our server, we analyzed shorter RNA sequences and observed better results as compared to other methods.

The organization of paper is as follows: Sect. 2 is Materials and Method, we describe the proposed method with all the parameters taken into consideration along with the algorithm, description of how to use Web portal and performance measurement. Section 3 is Results and Discussion, we compared our method with three other methods and evaluated the performances with original RNA secondary structure and described the results along with observations.

2 Materials and Method

More the stacking regions in RNA, more stable the structure is, where stacking region is the region between two base pairs. So, the requirement is to maximize the number of base pairs (Fig. 1).

To explain the method, say for a given RNA sequence 'AUCGCCGGU', we find all possible base pairs using a base pairing matrix [19] as shown in Fig. 2(i). The RNA sequence is taken as row and column header, for every possible base pair G:C, A:U, G:U, we mark 1 for intersection of base pairs in the matrix. G:C and A:U are known as Watson–Crick base pair, and G:U is known as non-Watson–Crick base pair. We take only the upper right triangular matrix for the subsequent steps, as the matrix generated is symmetric. Next we consider a circle graph as shown in Fig. 2(ii) with each nucleotides as its vertex and possible base pairs as chords. As stated in literature, for a stable structure, the minimum number of nucleotides in loop region is supposed to be least 3 [19]. Taking this constraint, we remove certain base pairs, where two bases can pair, if there are more than two bases between them as shown in Fig. 2(iii). Then, we map the circle graph to an adjacency graph as in Fig. 2(iv), and

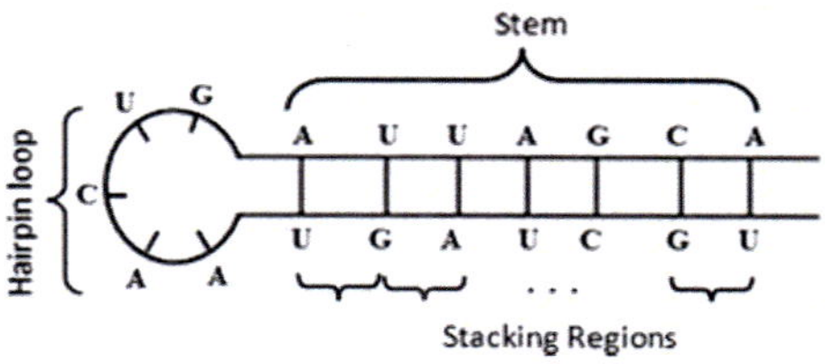

Fig. 1 RNA secondary structure with Stem and Hairpin loop

we take all chords of circle graph as new nodes of adjacency graph and intersecting chords of circle graph as new edges of adjacency graph.

For a chord say '2' between 'A' at node '1' and 'U' at node '9' as shown in Fig. 2(iii), two variables are taken as '*from*' and '*to*,' where '*from*' < '*to*,' hence, *from* of chord '2' is '1' (*from*(2) = 1), similarly *to* of chord '2' is '9' (*to*(2) = 9) as the chord '2' emerges from vertex 1 and ends at 9.

For intersection of chords, we check the following conditions taking every two chords say 'a' and 'b' in circle graph as follows: *from*(a) < *from*(b) < *to*(a) < *to*(b), *from*(b) < *to*(a) < *to*(b) < *to*(a), *to*(a) = *to*(b), *to*(a) = *from*(b), *from*(a) = *to*(b), and *from*(a) = *from*(b). The first two conditions check if two chords are intersecting. The last four checks if two chords have same vertex in common. As an example, in Fig. 2(iii) chord '2' and '11' are intersecting, as they have same vertex '9' in common, so in adjacency graph there will be an edge between '2' and '11'.

Next, we find all possible maximum independent sets (MIS) of the adjacency graph, here in case we have only one MIS {2,5,7}, which are dark circled as shown in Fig. 2(iv). In the next step, we choose the edges of MIS from the circle graph. So finally, we get a planar graph as shown in Fig. 2(v) by choosing the chords of circle graph named '2','5','7'.

In this example, we have only one MIS. But we may have multiple MIS for a given sequence. In that scenario, to resolve the conflict, first we choose the structures with maximum number of stacking regions. If still the conflict exists, then we check for structures having maximum consecutive stacks, if conflict still persists, then we compare the energies of stacks based on the stacking energy Table 1 [20]. If conflict still remains, we then compare the individual bond energies, along with loop energies also known as Tinoco's stability number [19] (Table 2).

2.1 Algorithm RNA Structure Estimation

Following is the list of functions and variables/constants used in the algorithm

1. **BPM**(rna_seq): for a given RNA sequence, it returns all possible base pairs
2. **CG**(Base_Mat): Maps the Base_Mat to a circle graph, where:
   Bases in row header of Base_Mat = Vertices aligned to circumference of circle graph
   Base pairs of Base_Mat = Chords of circle graph, joining two bases
3. **CHG**(Cir_Graph): Returns a circle graph, by keeping only those base pairs where distance between bases is more than two, called hairpin condition
4. **ADJ**(Cir_Hpin_Graph): Maps Cir_Hpin_Graph to an adjacency graph, where:
   Chords of Cir_Hpin_Graph = Vertices of Adjacency Graph
   Intersecting chords of Cir_Hpin_Graph = Edges of Adjacency Graph
5. **Largest_vertex_set**(Adj_Graph): This function returns maximum independent set (MIS) of adjacency graph, computed using Python igraph package. It returns a 2D matrix MIS_Mat, containing all possible sets of MIS. MIS_Mat[i] represent each 1D matrix, i.e., i^{th} row of MIS_Mat, where i ranges from 1 to no. of possible MIS.

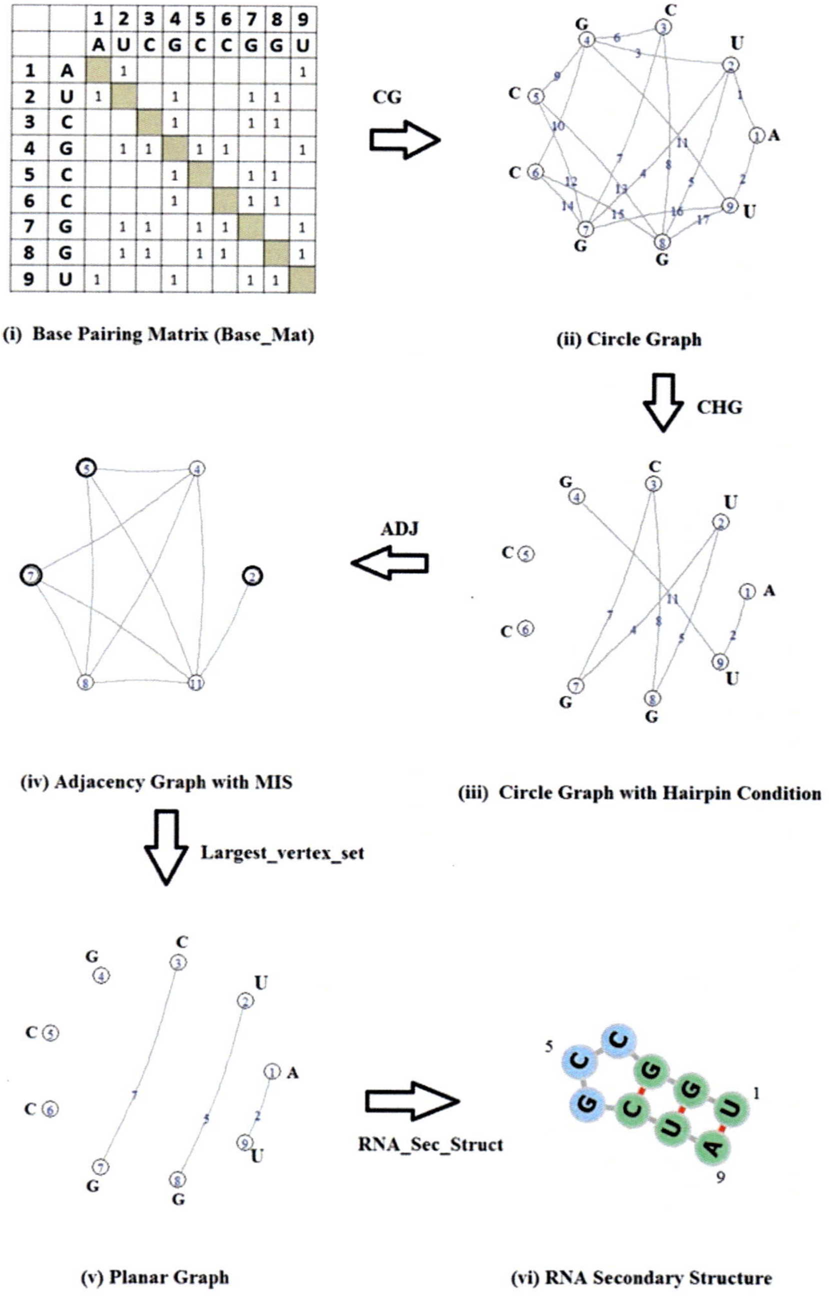

Fig. 2 Steps followed to detect RNA secondary structure

Table 1 Stacking energy Table 1 [20]

	A/U	C/G	G/C	U/A	G/U	U/G
A/U	−0.9	−1.8	−2.3	−1.1	−1.1	−0.8
C/G	−1.7	−2.9	−3.4	−2.3	−2.1	−1.4
G/C	−2.1	−2.0	−2.9	−1.8	−1.9	−1.2
U/A	−0.9	−1.7	−2.1	−0.9	−1.0	−0.5
G/U	−0.5	−1.2	−1.4	−0.8	−0.4	−0.2
U/G	−1.0	−1.9	−2.1	−1.1	−1.5	−0.4

The leftmost column represents current base pair and the topmost row represents next base pair. Data in row 2 column 1 represent the energy when C/G is followed by A/U

Table 2 Tinoco's stability number

Size	HP	BL	IL
<2	NA	−2	NA
2	NA	NA	−4
3	−5	−3	−5
4 to 7	−6	NA	−6
>7	−7		−7
4 to 15	NA	−5	NA
>15	NA	−6	NA

6. **CS**(MIS_Mat[i]): A stack consists of two consecutive base pairs. This function returns an array containing total number of stacks, in each MIS_Mat[i]
7. Max(Count_Stack): It returns the maximum number of stacks comparing each MIS_Mat[i], and the count of how many maximum values
8. **CQS**(MIS_Mat[i]): This function returns an array containing total number of highest consecutive stacks in each MIS_Mat[i]
9. Max(Count_Consqutv_Stack): It returns the maximum number of consecutive stacks comparing each MIS_Mat[i] and the count of how many maximum values
10. **SE**(MIS_Mat[i]): It returns the total stacking energy as per table given in stacking energy Table 1 [20]
11. Min(Stack_Energy): It returns minimum stacking energies comparing each MIS_Mat[i]
12. **TSN**(MIS_Mat[i]): It returns Tinoco's stability number comparing each MIS_Mat[i], for a particular MIS_Mat[i], TSN is the sum of base pair energy (BP), hairpin energy (HP), bulge loop energy (BL), and interior loop energy (IL)

$$TSN = \sum BP + \sum HP + \sum BL + \sum IL \quad (1)$$

The energies are as follows:

$$BP = \begin{Bmatrix} 1, \text{ for A:U base pair} \\ 2, \text{ for G:C base pair} \\ 0, \text{ for G:U base pair} \end{Bmatrix} \tag{2}$$

13. Max(Tinoco_Stability_No): It returns maximum stability number when compared to each MIS_Mat[i].
14. **RNA_Sec_Struct**(MIS_Mat[i]): This function represents MIS_mat[i] as a dot bracket notation, where brackets (or) represent nucleotides that participate in a base pair and . represents nucleotides that do not participate in base pair. From the dot_bracket notation, corresponding RNA secondary structure can be visualized.

```
Result: Estimated RNA secondary structure
Base_Mat = Compute BPM(rna_seq)
Cir_Graph = Compute CG(Base_Mat)
Cir_Hpin_Graph =Compute CHG(Cir_Graph)
Adj_Graph = Compute ADJ(Cir_Hpin_Graph)
MIS_Mat = Largest_vertex_set(Adj_Graph)
if length(MIS_Mat) = 1 then
 | Compute RNA_Sec_Struct(MIS_Mat)
else
 | Count_Stack = CS(MIS_Mat[i])
 | Count_Stack_Max = Max(Count_Stack)
 | if length(Count_Stack_Max) = 1 then
 |  | Compute RNA_Sec_Struct(MIS_Mat[i])
 | else
 |  | Count_Consqutv_Stack = CQS (MIS_Mat[i])
 |  | Count_Consqutv_Stack_Max = Max(Count_Consqutv_Stack)
 |  | if length(Count_Consqutv_Stack_Max) = 1 then
 |  |  | Compute RNA_Sec_Struct(MIS_Mat[i])
 |  | else
 |  |  | Stack_Energy = Compute SE(MIS_Mat[i])
 |  |  | SE_Min = Min(Stack_Energy)
 |  |  | if length(SE_Min) = 1 then
 |  |  |  | Compute RNA_Sec_Struct(MIS_Mat[i])
 |  |  | else
 |  |  |  | Tinoco_Stability_No = Compute TSN(MIS_Mat[i])
 |  |  |  | Tinoco_Stability_No_Max = Max(Tinoco_Stability_No)
 |  |  |  | for all MIS_Mat[i] with Tinoco_Stability_No_Max
 |  |  |  | Compute RNA_Sec_Struct(MIS_Mat[i])
 |  |  | end
 |  | end
 | end
end
```

The implementation of algorithm is available in a Web portal from Tezpur University (TU Web server), which is accessible at the link http://14.139.219.242:8003/rna_struct.

2.2 Description of How to Use TU Web Server

Step I: Enter nucleotide sequence of RNA: The first step is to provide nucleotide (base) sequence of RNA for which secondary structure is to be detected.

Step II: Enter the restrictions for each nucleotide (Optional): This step is optional, where the user is provided with an option to impose restriction on bases, of which to pair and which not to. For every base, a character 'x' is to be entered to restrict the base to pair, and a character '.' to allow the base to pair.

Step III: Select the base pairs to keep: In this step, we provide an option to the user, to select base pairs which are to be included in RNA structure detection.

Step IV: Enter e-mail id: This step is optional, an e-mail id can be provided, if the user wants the results in their e-mail.

In the next step, the user may hit the Calculate button to view the results, a dot bracket notation, circle graph of RNA structure and a link [21] to visualize the RNA structure is also provided.

2.3 Performance Measurement

To determine the accuracy of our method and other known methods as provided in Web servers of Vienna RNA fold [22, 23], RNAStructure [24, 25], Cofold [14, 26] in comparison to original RNA structures we perform sensitivity (SS), specificity (SP), and correlation coefficient measures as follows:

$$SS = \frac{TP}{TP + FN}, \quad SP = \frac{TP}{TP + FP}$$

$$CC = \sqrt{\left(\frac{TP}{TP + FN}\right) * \left(\frac{TP}{TP + FP}\right)}$$

where the confusion matrix is provided below, BP means base pair:

	BP predicted: No	BP predicted: Yes
BP exists: No	True Negative (TN)	False Positive (FP)
BP exists: Yes	False Negative (FN)	True Positive (TP)

3 Results and Discussion

For this study, RNAs which have been used in the literature are taken for comparison analysis. The first four RNAs as depicted in Table 3 have been collected from online database (http://server3.lpm.org.ru/urs/struct.py) named Universe of RNA structures, the sequence ID given are PDBId of RNA sequence, the method applied for analysis of these RNA structures are either NMR or X-ray crystallography and are considered as original RNA secondary structure. They also provide a dot bracket notation for each RNA structure, which aids in comparison of RNA structures from different computational sources [23, 25, 26]. The last two RNAs are taken from literature [3, 18, 27].

SS is the probability of correctly predicting base pairs, whereas SP is the probability that a base pair prediction is correct [28]. From the above table, we can say that in terms of sensitivity (SS) our algorithm has higher probability of predicting correct base pairs as SS is almost 1.0 in almost all cases, as compared to other methods, whereas the specificity (SP) measure is comparable of our Web server to other methods in the sequences of 3DKN, 1RAW, and 1F1T. In our method, the correlation coefficient measure performs better in case of 2JTP, 3DKN, E_coli_16S_rRNA and R17_Viral_RNA, but other methods perform well in case of 1F1T, 1RAW.

In this study, we proposed a method that determines the maximum possible base pairs in an RNA secondary structure with no intersections. We used the maximum independent set (MIS) approach, that gives all possible combinations of base pairs that are maximum in number. The computational time complexity is O(nmμ), where n is the number of vertices, m number of edges, and μ number of maximum independent sets of the circle graph [29]. Our proposed method not only maximizes the number of base pairs, but stacking regions also, as it is known that more the stacks in a RNA secondary structure, more stable the structure will be.

It has been seen that, in small RNAs not having bifurcation, maximum number of base pairs are possible when the first bases pair with last bases of a RNA sequence, thus reducing the number of base pairs and still predicting the correct secondary structure. In our implementation, we took some threshold values.

Considering two position of bases p and q and length of RNA sequence as l, we choose base pairs (bp) with the following conditions:

$$bp = \left\{ \begin{array}{ll} (l-3) < (p+q) < (l+3), & l < 10 \\ (l-6) < (p+q) < (l+6), & l < 400 \\ (l-10) < (p+q) < (l+10), & l > 400 \end{array} \right\} \quad (3)$$

We observed that, number of MIS generated is independent of sequence length. However, large number of MIS might be generated for a sequence rich in AT or GC bases, possibly because they lead to large number of base pairs.

Based on features considered, it is difficult to decide which method is better compared to which other and therefore the methods can be considered as complimentary to each other.

Table 3 Comparative result

Sequence name	L	Vienna RNAfold Web server			RNAStructure Web server			Cofold Web server			TU Web server		
		SS	SP	CC	SS	SP	CC	SS	SP	CC	SS	SP	CC
1F1T	38	1.00	1.00	100.00	1.00	1.00	100.00	1.00	1.00	100.00	**1.00**	0.93	96.36
1RAW	36	0.75	0.75	75.00	0.75	0.75	75.00	0.75	0.75	75.00	**0.80**	0.67	73.03
2JTP	34	1.00	1.00	100.00	0.85	1.00	91.99	1.00	1.00	100.00	**1.00**	1.00	**100.00**
3DKN	32	0.75	0.75	75.00	0.75	0.75	75.00	0.75	0.86	80.18	**1.00**	0.73	**85.28**
E_coli_16S_rRNA	38	0.77	0.77	76.92	0.77	0.77	76.92	0.92	1.00	96.08	**1.00**	0.87	**93.09**
R17_Viral_RNA	55	0.90	1.00	95.12	0.90	1.00	95.12	1.00	1.00	100.00	**1.00**	1.00	**100.00**

L sequence length (number of bases); *SS* sensitivity; *SP* specificity; *CC* correlation coefficient

References

1. Qasim, R., Kauser, N., Jilani, T.: Secondary structure prediction of RNA using machine learning method. Int. J. Comput. Appl. **975**, 8887 (2010)
2. Ji, Y., Xu, X., Stormo, G.D.: A graph theoretical approach for predicting common RNA secondary structure motifs including pseudoknots in unaligned sequences. Bioinformatics **20**(10), 1591–1602 (2004)
3. Takefuji, Y., Chen, L.: Parallel algorithms for finding a near-maximum independent set of. IEEE Trans. Neural Networks **1**(3), 263 (1990)
4. Bernhart, S.H., Hofacker, I.L., Will, S., Gruber, A.R., Stadler, P.F.: RNAalifold: improved consensus structure prediction for RNA alignments. BMC Bioinform. **9**(1), 474 (2008)
5. Bonnet, E., Rzewski, P., Sikora, F.: Designing RNA secondary structures is hard. J. Comput. Biol. (2020)
6. Hofacker, I.L., Stadler, P.F.: Automatic detection of conserved base pairing patterns in RNA virus genomes. Comput. Chem. **23**(3–4), 401–414 (1999)
7. Wang, L., Liu, Y., Zhong, X., Liu, H., Lu, C., Li, C., Zhang, H.: DMfold: a novel method to predict RNA secondary structure with pseudoknots based on deep learning and improved base pair maximization principle. Front. Genet. **10**, 143 (2019)
8. Zuker, M.: Prediction of RNA secondary structure by energy minimization. In: Computer Analysis of Sequence Data, pp. 267–294. Springer, Berlin (1994)
9. Zuker, M.: Mfold web server for nucleic acid folding and hybridization prediction. Nucl. Acids Res. **31**(13), 3406–3415 (2003)
10. Zuker, M., Stiegler, P.: Optimal computer folding of large RNA sequences using thermodynamics and auxiliary information. Nucl. Acids Res. **9**(1), 133–148 (1981)
11. Knudsen, B., Hein, J.: Pfold: RNA secondary structure prediction using stochastic context-free grammars. Nucl. Acids Res. **31**(13), 3423–3428 (2003)
12. Van Batenburg, F., Gultyaev, A.P., Pleij, C.W.: An APL-programmed genetic algorithm for the prediction of RNA secondary structure. J. Theor. Biol. **174**(3), 269–280 (1995)
13. Li, J., Xu, C., Wang, L., Liang, H., Feng, W., Cai, Z., Wang, Y., Cong, W., Liu, Y.: PSRNA: prediction of small RNA secondary structures based on reverse complementary folding method. J. Bioinform. Comput. Biol. **14**(04), 1643001 (2016)
14. Proctor, J.R., Meyer, I.M.: Cofold: thermodynamic RNA structure prediction with a kinetic twist. arXiv preprint arXiv:1207.6013 (2012)
15. Garey, M.R., Johnson, D.S.: Computers and Intractability, vol. 29. W. H. Freeman, New York (2002)
16. Gavril, F.: Algorithms on circular-arc graphs. Networks **4**(4), 357–369 (1974)
17. Hsu, W.L.: The coloring and maximum independent set problems on planar perfect graphs. J. ACM (JACM) **35**(3), 535–563 (1988)
18. Liu, Q., Ye, X., Zhang, Y.: A Hopfield neural network based algorithm for RNA secondary structure prediction. In: First International Multi-Symposiums on Computer and Computational Sciences (IMSCCS'06), vol. 1, pp. 10–16. IEEE, New York (2006)
19. Tinoco, I., Uhlenbeck, O.C., Levine, M.D.: Estimation of secondary structure in ribonucleic acids. Nature **230**(5293), 362–367 (1971)
20. Turner, D.H., Sugimoto, N., Freier, S.M.: RNA structure prediction. Ann. Rev. Biophys. Biophys. Chem. **17**(1), 167–192 (1988)
21. Kerpedjiev, P., Hammer, S., Hofacker, I.L.: Forna (force-directed RNA): simple and effective online RNA secondary structure diagrams. Bioinformatics **31**(20), 3377–3379 (2015)
22. Lorenz, R., Bernhart, S.H., Zu Siederdissen, C.H., Tafer, H., Flamm, C., Stadler, P.F., Hofacker, I.L.: ViennaRNA package 2.0. Algorithms Mol. Biol. **6**(1), 26 (2011)
23. Vienna RNAfold: http://rna.tbi.univie.ac.at//cgi-bin/RNAWebSuite/RNAfold.cgi
24. Reuter, J.S., Mathews, D.H.: RNAstructure: software for RNA secondary structure prediction and analysis. BMC Bioinform. **11**(1), 129 (2010)
25. RNAstructure: https://rna.urmc.rochester.edu/RNAstructureWeb/Servers/Predict1/Predict1.html

26. CoFold: https://www.e-rna.org/cofold/cite.cgi
27. Stern, S., Weiser, B., Noller, H.F.: Model for the three-dimensional folding of 16 S ribosomal RNA. J. Mol. Biol. **204**(2), 447–481 (1988)
28. Baldi, P., Brunak, S., Chauvin, Y., Andersen, C.A., Nielsen, H.: Assessing the accuracy of prediction algorithms for classification: an overview. Bioinformatics **16**(5), 412–424 (2000)
29. Tsukiyama, S., Ide, M., Ariyoshi, H., Shirakawa, I.: A new algorithm for generating all the maximal independent sets. SIAM J. Comput. **6**(3), 505–517 (1977)

NTP Server Clock Adjustment with Chrony

Amina Elbatoul Dinar, Boualem Merabet, and Samir Ghouali

Abstract As of now, all servers have an equipment or programming clock to which reference is made to time stamp records, exchanges, messages, and so forth. This clock, albeit structured around a quartz oscillator, floats like any customary watch, which implies this common watch cannot a match to such created machines that are networked and share common resources like file systems. For example, UNIX is a development tool which makes command, based on its work on comparing file modification dates. Similarly, the correlation of log messages from several systems becomes very difficult if they are not at the same time. In this article, we will concentrate on this topic by designing a server utilizing the NTP convention since the primary "focus" of the NTP usage is UNIX frameworks, and to be more explicit, we will see the management of the NTP server with the Chrony tool.

Keywords Network time protocol · Servers synchronization · Chrony · Kali linux

1 Introduction

On servers, numerous procedures use time [1–4], some record the hour of a client's association in a log, others the hour of a request for an online deals framework for instance. Time exactness turns out to be especially basic when a few machines cooperate; they need a period estimation to synchronize their activities.

A. E. Dinar (✉) · B. Merabet · S. Ghouali
Faculty of Sciences and Technology, Mustapha Stambouli University, Mascara, Algeria
e-mail: amina.dinar@univ-mascara.dz

B. Merabet
e-mail: boualem19985@yahoo.fr

S. Ghouali
e-mail: s.ghouali@univ-mascara.dz

A. E. Dinar
LSTE Laboratory, University Mustapha Stambouli of Mascara, Mascara, Algeria

S. Ghouali
STIC Laboratory, Faculty of Engineering, University of Tlemcen, Tlemcen, Algeria

J. K. Mandal et al. (eds.), *Applications of Internet of Things*, Lecture Notes in Networks and Systems 137, https://doi.org/10.1007/978-981-15-6198-6_16

Companies in the transport sector also have a major interest in supporting their computers systems and networks with servers using NTP and PTP protocols, particularly to ensure more efficient use of their GPS. For an aircraft, flying at an average speed of nearly 1000 km/h, a one-second delay represents a position error of more than 250 m. An hourly reliability including all parameters (zones leap years… etc.) becomes essential.

The high time resolution obtained allows computer and/or robotic accuracy at a scale exceeding one millisecond and thus allows for greater efficiency and production speed, thanks to the coordination of the machines. In this way, the sequencing of these exercises therefore gains in robotization, and the groups working with these machines are then increasingly effective.

For health care facilities, a time synchronization system is particularly important to: ensure proper planning of medical teams; proper administration of medication at the right time and in the right order of prescription; ensure the smooth running of surgical procedures. Datacenters need a time domain in the millisecond range for platform virtualization. The chronology of events also allows errors to be traced on the same millisecond scale: Traceability ensures a backup, or automatic backup, at night requiring an accuracy of about ten seconds. This increases the reliability of daily backups; the time server allows protecting against time deviations caused by an electrical frequency that is not stable enough which varies permanently around 50 Hz in Europe, and the synchronization provided by the server in NTP allows reliable and robust clustering [5–7].

This paper is composed as follows: Sect. 2 presents distinctive synchronization convention framework systems. We focused our study on NTP configurations and variety of reference clocks and sources. Section 3 consists to administrate NTP by Chrony tool to which its task are compared with those of NTP on Sect. 4. Section 5 presents security NTP mechanism and attack detection to ensure it legitimacy and trustworthiness. Finally, Sect. 6 concludes the paper with future directives.

2 How to Ensure the Synchronization of Networked Equipment?

2.1 Time Protocol

It is the subject of RFC868; relying on UDP or TCP, it can be summarized as the servers sending a packet containing the time in seconds elapsed since January 1, 1900, at 0H. Time protocol was used by the UNIX timed daemon but its low resolution and the lack of specification of transit time compensation mechanisms led to the study of a more sophisticated protocol [8].

2.2 Simple Network Time Protocol (SNTP)

SNTP (SNTPv4) is proposed for essential servers furnished with a solitary reference clock, just as for customers with a solitary upstream server and no reliant customers. The completely created NTPv4 usage is expected for optional servers with various upstream servers and numerous downstream servers or customers. Other than these contemplations, NTP and SNTP servers and customers are totally interoperable and can be intermixed in NTP subnets. A SNTP essential server executing the on-wire convention has no upstream servers with the exception of a solitary reference clock. On a basic level, it is undefined from a NTP essential server that has the alleviation calculations and accordingly fit for moderating between various references tickers [9].

2.3 Network Time Protocol (NTP)

It is the subject of RFC1305 and is in its third version. Much more elaborate than time protocol, it allows the creation of networks of NTP entities with multiple redundancies in order to ensure the permanent and reliable synchronization of the machines concerned. The main contribution to the work on NTP is that of D. L. Mills from the University of Delaware [3]. Filtering and selection algorithms and implementation models are defined in NTP. They allow NTP clients to determine the best source of synchronization, eliminate suspicious sources, and correct network transit times at any time. Regarding its implementation, one of the main characteristics of an NTP network is its pyramidal structure [10]. Time references synchronize NTP servers that are directly connected to them. These constitute "stratum" 1, and they will each synchronize several dozen other servers that will constitute "stratum" 2 and so on up to the terminal clients. This principle makes it possible to distribute the load of the servers well while maintaining a "distance" to the relatively small reference sources [11, 12].

NTP is therefore a protocol that allows synchronizing the time of different systems through an IP network. Clients synchronize their clocks with servers. These servers synchronize themselves with other servers and so on. This network is organized in layers called stratums [5, 13].

The network time protocol (NTP): Presented in 1985 as RFC 958 by D. L. Mills and modified in 2010 in form NTPv4 as RFC 5905, the network time protocol is a long standing and wide-spread convention for appropriating time data. NTP utilizes the association—less UDP convention by means of port 123. Its engineering works with a progressively layered correspondence model. Getting time data from stratum 0 sources stratum 1 servers convey an opportunity to layers beneath, etc. With each layer, the stratum number increases, and the feasible precision diminishes. Other than other correspondence models, the unicast mode (customer to server, server to customer) is the most predominant usual way of doing things.

2.4 NTP Configuration

This segment (Section) gives best practices to NTP arrangement and activity. Application of these accepted procedures that are explicit to the network time foundation implementation.

2.4.1 Staying up with the Latest

There are numerous renditions of the NTP convention being used, and various usages on a wide range of stages. The practices right now intended to apply by and large to any execution of RFC5905. NTP clients should choose a usage that is effectively kept up. Clients should stay up with the latest on any known assaults on their chose execution and send refreshes containing security fixes when pragmatic.

2.4.2 Utilize Enough Time Sources

A NTP execution that is consistent with [RFC5905] takes the accessible wellsprings of time and presents this planning information to advanced crossing point, grouping, and joining calculations to get the best gauge of the right time.

2.4.3 Utilize an Assorted Variety of Reference Clocks

When utilizing servers with appended equipment reference timekeepers, it is proposed that various kinds of reference tickers be utilized. Having sources with autonomous executions implies that any one issue is more averse to cause assistance interference [8].

An NTP server can operate in the following modes:

- Simple server mode: it only responds to requests from its clients.
- Active symmetric mode: it asks to be synchronized by other servers and announces to them that it can also synchronize them.
- Passive symmetric mode: same thing but on the initiative of other servers.
- Broadcast mode: intended for local networks, it is limited to the distribution of time information to customers who may be either passive or discover the servers with which they will synchronize.
- Client mode: sends requests to one or more servers.

To synchronize our clocks with our computer network, the most secure and dependable strategy is to have a committed NTP or SNTP server. The architecture in NTPv4 allows a 10× greater time accuracy than the old NTPv3 protocol [14, 15]. The proximity (of the server to the network) provides a minimum latency between the server and our clocks, computers and other equipment.

The implementation of the NTP protocol as well as various drivers used for the connection of time references permit implementing both a simple terminal client and a primary server. The purely NTP part runs on a large number of operating systems: Solaris 2, HP/UX 9.x, SunOS 4.x, OSF/1, IRIX 4.x, Ultrix 4.3, AIX 3.2, A/UX, BSD, Kali Linux. Achieving good accuracy depends on how well the messages are identified at:

The application level UNIX is not a real-time system, and it is the least efficient solution but the easiest to implement.

The level of the kernel software queues much more precise solution but requires intervention in the kernel.

In our study, our operating system is Linux (Kali), we configure the time of our machine and set the system time with timedatectl, and this command will display the time information of our system:

```
root@amina - kali: - # timedatectl
Local time: mar. 2019 - 08 - 20 23:33:36 CET
Universal time: mar. 2019 - 08 - 20 22:33:36 UTC
RTC time: mar. 2019 - 08 - 20 23:33:36
Time zone: Africa/Algiers (CET, +0100)
System clock synchronized: yes
NTP service: inactive
RTC in local TZ: yes
Use RTC in UTC by calling 'timsdatectl set - local - rtc 0'.
```

If the clock is not automatically synchronized online, the server time can be configured using set-time: **#sudo timedatectl set - time**.

We list the different time zones by list-timezones: **#timedatectl list - timezones |grep Algeria**.

The time zone is configured using set-timezone: **# sudo timedatectl set - timezoneAfrica/Algeria**.

One of the largest clusters of public NTP servers is called **pool.ntp.org.**. This one is configured by default in most Linux distributions.

Under the latest versions of Linux, the system clock is automatically synchronized in a network. This synchronization is managed by the systemd-timesyncd.service service. More information about this service can be accessed by the command: **# Systemctl status systemd - timesyncd**.

$ timedatectl

It is therefore possible to synchronize the clock of all the servers on your network by synchronizing each of them with the global NTP network, but as soon as the network grows, it becomes advantageous to have your own NTP server.

There are several other NTP concepts: stepping, slewing, insane time, drift, and jitter.

```
remote            refid                  st t  when poll reach  delay   offset  jitter
==============================================================================
0.kali.pool.n .     POOL.                 16 p  -    64  0     0.000   0.000  0.000
1.kali.pool.n .     POOL.                 16 p  -    64  0     0.000   0.000  0.000
2.kali.pool.n .     POOL.                 16 p  -    64  0     0.000   0.000  0.000
3.kali.pool.n .     POOL.                 16 p  -    64  0     0.000   0.000  0.000
ntp.kali.com .      POOL.                 16 p  -    64  0     0.000   0.000  0.000
-102.130.49.223     85.199.214.98          2 u 792 1024 337  242.941 -22.544  4.332
-ns.bitco.co.za     41.78.128.17           3 u 759 1024 377  242.113 -19.735  4.662
-160.119.238.133    196.21.187.2           2 u 786 1024 377  260.651 -21.193  8.842
+chilipepper.can    145.238.203.14         2 u 779 1024 377   91.593 -24.107  2.746
*pugot.canonical    17.253.108.253         2 u 928 1024 377   85.897 -30.369  7.252
-dbn-ntp.mweb.com   194.58.204.148         2 u 791 1024 377  287.697 -20.849  4.199
+golem.canonical    145.238.203.14         2 u 311 1024 377   93.789 -32.492  6.703
```

- Remote: specifies the hostname address of time provider that is we are getting time from we have,
- Refid: indicates the type of time reference source that we are connecting to,
- st: specifies the stratum of that time provider,
- When: specifies the number of seconds since the last time poll occurred,
- Poll: indicates the number or seconds between tow time polls,
- Reach: is the key to knowing that NTP is working properly because it is a circular bit buffer, it show us the statue of the last eight NTP messages (377 is the eight octal bit). Each NTP missed packet response is tracked over in the next eight NTP update intervals reach field,
- Offset: specifies the time difference between the local system and the time on the time provider which is in milliseconds [16].

#ntptrace
Localhost: stratum 3, offset −0.046775, synch distance 0.152070

We use also the Ntptrace command to monitor time synchronization, he specifies the time provider's stratum which lists also the time offset between the local system and the time provider.

Indeed, having your own NTP server allows you to: improve synchronization between network servers, reduce traffic due to time synchronizations on the Internet connection, keep servers synchronized even in the event of an Internet outage, and avoid unnecessary strain on the global NTP network.

3 NTP Server with Chrony

Kali Linux uses Chrony software as the default NTP server, and this program is installed by the command: **#sudo apt-get install chrony**.

Then, we configure Chrony by editing the file/etc./chrony/chrony.conf. In this configuration file, there is a certain amount of information, such as: The line beginning with pool indicates the address of the NTP servers (or groups of servers more precisely) to be used and the maximum number of resources to be used. A priori, we can continue to use the default selection.

Drift file indicates the file to use to record the time drift of the server from the pool. It allows you to resynchronize the clock faster. Chrony does not allow customers to synchronize with this time service. The clients' network must be authorized by allowing directive by editing the following line at the end of the file: The address of our network, for example allow 192.168.0/24.

We can launch Chrony and activate it when the server starts: **#sudo systemctl enablechrony** and **# sudo systemctl start chrony**.

Chrony listens on UDP port 123 (default port for the NTP service).

Make sure that this port on the firewall was opened, so that clients can synchronize.

As Chrony is now in charge of synchronizing our system clock, we disable systemd-timesyncd by: **$ sudo timedatectl set-ntp false**.

Chrony provides a command line interface to query and manage Chrony: chronyc. We can therefore display the servers with which we are synchronized by the command:

$ chronyc sources.

The server that starts with ^* is the current time source. Those starting with ^+ are used to calculate an average time, and those starting with ^– are not currently used.

4 NTP Chrony Comparison Tasks

NTP underpins the auto key convention to validate servers with open key cryptography. Note that the convention has been demonstrated to be unreliable, and it will be presumably supplanted with a usage of the network time security (NTS) particular, NTP has been ported to even more working frameworks, he incorporates an enormous number of drivers for different equipment reference timekeepers, chrony requires different projects (for example gpsd or ntp-refclock) to give reference time by means of the SHM or SOCK interface, and he can perform helpfully in a situation where access to time reference is irregular. NTP needs normal surveying of the reference to function better, and he can as a rule synchronize the clock quicker and with more time precision. It rapidly adjusts to unexpected changes clock (for example because of changes in the temperature of the precious stone oscillator), and he can perform well not withstanding when the system is clogged for longer timeframes.

Chrony bolsters equipment time stamping on Linux, which permits very exact synchronization on neighborhood systems, and he offers help to work out the addition or misfortune pace of the continuous clock, for example, the clock that keeps up when the PC is killed. It can utilize this information when the framework boots to set the framework time from a redressed adaptation of the ongoing clock. These continuous clock offices are just accessible on Linux, up until now [14].

5 NTP Security Mechanisms

In the standard arrangement, NTP groups are exchanged unprotected among client and server. A foe that can turn into a man-in-the-middle is subsequently ready to drop, replay, or change the substance of the NTP parcel, which prompts debasement of the time synchronization or the transmission of bogus time data. A hazard assessment for time synchronization is given in [RFC7384]. NTP gives two inner security systems to ensure legitimacy and trustworthiness of the NTP parcels. The two measures ensure the NTP parcel by methods for a message authentication code (MAC). Neither of them scrambles the NTP's payload, since this payload data is not viewed as secret. Detection of attacks [17] through monitoring administrators should screen their NTP instances to identify assaults. Many known assaults on NTP have specific marks. Ordinary attacks marks include:

1. Zero root parcels: a bundle with a source timestamp set to zero.
2. A bundle with an invalid cryptographic MAC.

The perception of numerous such bundles could show that the customer is enduring an onslaught [18].

6 Conclusion, Perspectives and Some Advices

In this article, we concentrate on this subject by designing a server utilizing the NTP convention since the primary focus of the NTP execution is UNIX frameworks, to be progressively unequivocal; we see the administration of the NTP server with the Chrony tool. On a local network, the use of broadcast mode makes it possible to simplify the configuration of clients. Distribute the load well by setting up as many layers as necessary, in particular so as not to overload the public reference servers. Soon, we will use versions of xntpd, only the redacted versions of the DES are exportable from the US, and they carry the word export in their name and are sometimes several numbers late compared to the current version. As xntpd continues to evolve rapidly, our research has led us to study in the near future, how to use UNIX implementation to secure the NTP server by Chrony.

References

1. Dinar, A.E., Ghouali., S., Merabet, B., Feham, M.: Packet synchronization in an network time protocol server and ASTM Elycsys packets during detection for cancer with optical DNA Biochip. In: International Congress on Health Sciences and Medical Technologies, Tlemcen, Algeria, 5–7 December (2019)
2. Zhao, K.J., Zhang, A.I., Mning, D.Y.: Implementation of network time server system based on NTP. Electronic Test **7**, 13–16 (2008)
3. Li, X.Z.H.: Research on the Network Time Synchronization System Based on IEEE1588. National Time Service Center, Chinese Academy of Sciences (2011)
4. Novick, A., Lombardi, M.: A comparison of NTP servers connected to the same reference clock and the same network. In: Proceedings of the 2017 Precise Time and Time Interval Systems and Applications Meeting, Monterey, California, pp. 264–270, 30 January–2 February (2017)
5. Warrington, R.B., Fisk, P.T.H., Wouters, M.J., Lawn, M.A., Thorn, J.S., Quigg, S., Gajaweera, A., Park, S.J.: Time and Frequency Activities at the National Measurement Institute, Australia. Frequency Control Symposium and Exposition. In: Proceedings of the IEEE International, pp. 231–234 (2005)
6. Mills, D.L.: Internet Time Synchronization: The Network Time Protocol. IEEE Trans. Commun. **39**(10) (1991)
7. IEEE Std 1588-2008: IEEE Standard for a Precision Clock Synchronization Protocol for Networked Measurement and Control Systems. IEEE1588-2008 Standard (2008)
8. https://www.thegeekdiary.com/what-is-the-refid-in-ntpq-p-output/. Last accessed 12 Aug 2019
9. Langer, M., Behn, T., Bermbach, R.: Securing Unprotected NTP Implementations Using an NTS Daemon. In: IEEE International Symposium on Precision Clock Synchronization for Measurement Control and Communication (ISPCS) (2019). https://doi.org/10.1109/ispcs.2019.8886645
10. Lombardi, M., Levine, J., Lopez, J., Jimenez, F., Bernard, J., Gertsvolf, M., et al.: International Comparisons of Network Time protocol Servers. In: Proceedings of the 2014 Precise Time and Time Interval Systems and Applications Meeting, Boston, Massachusetts, pp. 57–66, 1–4 December (2014)
11. Sommars, S.E.: Challenges in Time Transfer Using the Network Time Protocol (NTP). In: Proceedings of the 48th Annual Precise Time and Time Interval Systems and Applications Meeting, California, pp. 271–290, January (2017)
12. Vijayalayan, K., Veitch, D.: Rot at the roots examining public timing infrastructure. In: Proceedings of the 35th Annual IEEE International Conference on Computer Communications, San Francisco, California, pp. 1–9, April (2016)
13. Matsakis, D.: Time and Frequency Activities at the U.S. Naval Observatory. Frequency Control Symposium and Exposition. In: Proceedings of the 2005 IEEE International, pp. 271–224 (2005)
14. Mills, D.L.: RFC1305 - NTPv3. http://rfc-editor.org/. Last accessed 25 Oct 2018
15. Mills, D.L.: RFC4330 - SNTPv4. http://rfc-editor.org/. Last accessed 25 Oct 2018
16. https://chrony.tuxfamily.org. Last accessed 24 Sept 2018
17. Bennabti, S., Dinar, A.E., Merzougui, R., Merabet, B., Ghouali, S.: Risk cryptography planning in telecommunications systems 'CRYP-TS': attack strategy & ethical hacking. In: Conference on Electrical Engineering CEE, Ecole Militaire Polytechnique, Algiers (2019)
18. Hoffmann, M., Toorop, W.: NTP Working Group A. Malhotra Internet-Draft Boston University Intended Status: Informational K. Teichel Expires: 9 January 2020 PTB. https://datatracker.ietf.org/meeting/105/agenda/ntp-drafts.pdf. Last accessed 02 July 2019

Angle-Based Feature Extraction Method for Fingers of Hand Gesture Recognition

Mampi Devi and Alak Roy

Abstract In this paper, two types of features 'angle' feature and *Finger_Tips* distance feature extraction methods for gestures of finger recognition are proposed. The entire image is segmented into several spatial modules and the task of feature extraction is carried out on finger of the hand images. Application of this method is extended to medical systems, sign languages for hearing-impaired people, crisis management and disaster relief, entertainment and human- -robot interaction. This method is tested on medial axis transformation (MAT) image and it does not require any gloves for recognition. This feature extraction algorithm has an advantage of very low feature dimension.

Keywords Feature extraction · Classification · MAT image · Hand gestures recognition

1 Introduction

Shape is an important vision-based features used to describe the image contents. Extraction of shape feature from two-dimensional images of three- dimensional objects is a difficult task due to the information loss incurred in projecting an object from three-dimension to two-dimension. The task becomes more complicated when the images are corrupted with noise, distortion and occlusion. The various features like moments, curvature, spectral features can be used to describe the shape of an object. Many shape-based features are available in the literature. These shape-based

M. Devi
Department of Computer Science and Engineering, Tripura University, Suryamaninagar, Agartala, Tripura 799022, India
e-mail: drammpidevi@gmail.com

A. Roy (✉)
Department of Information Technology, Tripura University, Suryamaninagar, Agartala, Tripura 799022, India
e-mail: alakroy@tripurauniv.in

J. K. Mandal et al. (eds.), *Applications of Internet of Things*, Lecture Notes in Networks and Systems 137, https://doi.org/10.1007/978-981-15-6198-6_17

features represent the whole image or sometimes boundary image. If the features represent the whole image, then they are contour-based features and otherwise they are region-based features. Again the contour-based features are sub-categorized into global features and structural features. The global features are represented by segments or sections and structural features are represented as a whole. The global shape descriptors are area, perimeter, eccentricity, major axis length, minor axis length, convexity, principle axis, circular variance and elliptic variance. These global features are described the boundary shape. Again, the structural feature extraction methods such as chain code, polygon decomposition, smooth curve decomposition, scale space methods and syntactic analysis are capable of partial matching and unable to capture the global information. Since these global features are only described the boundary shape and structural features are unable to capture global information. Feature extraction is considered to be the most critical stage and plays a major role in the success of all image processing and pattern recognition systems. Accordingly, many sophisticated feature extraction techniques have been developed in the literature of document image analysis to deal with documents. This paper presents a very simple and efficient methods for extraction of two features of fingers for recognizing of hand gesture are proposed.

2 Proposed Method

The steps involved in the feature extraction method are depicted in the basic framework as in Fig. 1. Each step of a framework has been explained briefly in the subsequent sub-sections.

The steps involved to convert the captured RGB image to gray and binary image has done in pre-processing step. The task performed in the pre-processing phase is described in the previous paper [1].

2.1 Medial Axis Transformation (MAT)

The medial axis transformation (MAT) finds out the closest boundary points for each point in an object and finally gives the skeletal of the images. Here, MAT are used to convert the binary images of single-hand gestures to skeletal images. The steps to convert the binary image to MAT image are described in our previous paper [2].

2.2 Proposed Features

In this section, it explains how the proposed features angle between two fingers and $Finger\_Tips$ distance are extracted from the medial axis transformation (MAT)

Fig. 1 Proposed conceptual framework

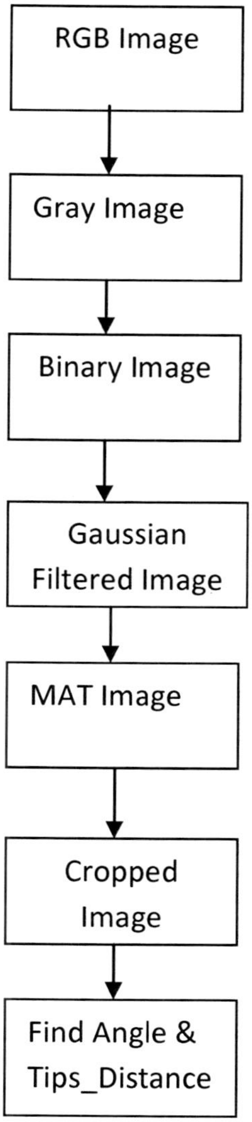

image. These two features are under the category of vision-based features which represents the shape of objects (hands) in a scene and can be visualized by normal eye. The features are invariant with respect to rotation and scaling. The method to extract the proposed features are presented by a simple mathematical formula. These features are very simple, however, very useful features for hand gestures recognition. Since the algorithm is used for fully open hand gestures, so it can applied the formula on the straight fingers. If the fingers are widely open, then it becomes easy to recognize the fingers. However, if the fingers are very close to each other, the hand images to MAT image conversion are different. For little gaps between the fingers of the hand, a different threshold is used to make the fingers more distinguishable. The threshold value is experimentally determined from a range of possible values. In these experiments, the best result is observed by taking the value is $1.9 \times graythreshold$. The extracted methods to find the two types of features are discussed in the following subsection.

2.3 Angle Between Fingers Feature

The feature 'angle between fingers' is used to measure the gap exists between fingers. Let x, y are the length of two fingers of a MAT image of a hand gestures as shown in the second image of Fig. 2. The angle A made by the two fingers of length x and y is given by the Eq. 1

$$z^2 = x^2 + y^2 - 2xy\cos A \tag{1}$$

$$\cos A = \frac{x^2 + y^2 - z^2}{2xy} \tag{2}$$

$$A = \cos^{-1}\left(\frac{x^2 + y^2 - z^2}{2xy}\right) \tag{3}$$

2.4 *Finger_Tips* Distance

Let p_1 and p_2 are the end points of first finger as shown in the second image of Fig. 2, p_1 and p_3 are the end points of second finger. Here, p_2 and p_3 are the tips of the both fingers. Let z is the distance between the finger tips.

And the distance between the $Finger\_Tips$, i.e., z can be approximated by the length of the arc between p_2 and p_3 made by the circle centered at p_1 with radius x

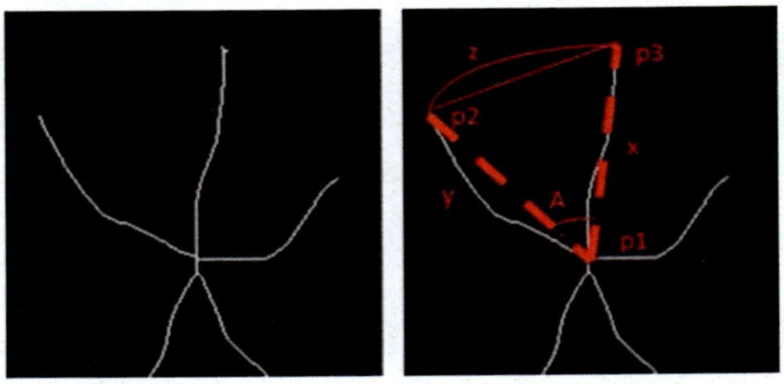

Fig. 2 Angle between fingers and *Finger_Tips* distance

(or y) as given in the following expression:

$$Finger_{Tips\ Distance} = \frac{3.14 * angle * radius}{180^\circ} \tag{4}$$

and x (or y)=$\sqrt{(p1 - p2)^2}$.

3 Experimental Result

The experimental description and the obtained outcome for the proposed method on single-hand gestures are discussed in the following sub-sections.

3.1 *Experimental Setup*

The experiment is carried out in a machine with configurations: Windows10 OS (64 bits), 4 GB RAM, 500 GB Hard disk and MATLAB 2015.

3.2 *Results Discussion*

The procedure to apply the proposed method is followed the following steps:
The above algorithm is implemented on single-hand gestures image using MATLAB

Algorithm 1: Find angle and $Finger\_Tips$ Distance

Input: Gray Image
Output: Angle, $Finger\_TipsDistance$
1. Convert gray image to MAT image.
2. Search row wise to find the x-coordinates of the junction points.
3. Similarly find the y-coordinates of the junction points
4. Print all the junction point
5. Find the angle between fingers using cosine mathematical formula.
6. Find the $Finger\_Tips$ Distance using above formula
7. Return angle and $Finger\_Tips$ Distance.

---

programming language. The output of the algorithm is shown in Fig. 3.

Fig. 3 Output results

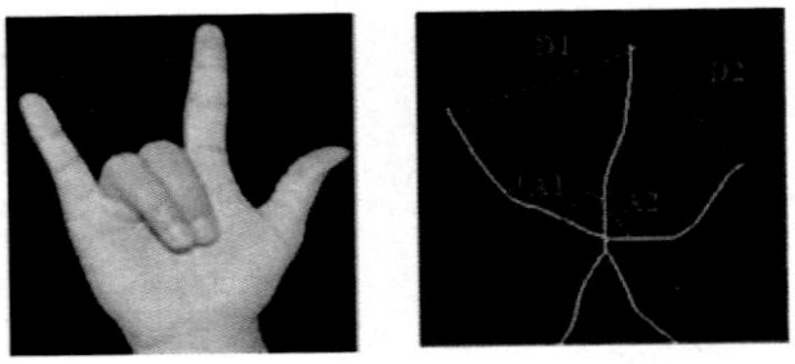

Input Converted MAT image

Result: Anglefinger_tip

row = 25

colm = 117

Angle(A1) = 90

Radius 1= 7.2801

Finger_Tips Distance (D1)= 11.4298

row = 137

colm = 139

Angle(A2) = 28.0725

Radius2 = 62.2415

Finger_Tips Distance (D2) = 30.4802

4 Conclusion and Future Direction

In this paper, two types of vision-based feature for fingers to recognize of hand gestures are extracted. The two proposed features are very simple but it has wide applications. The methods and algorithm to extracted features are explained in this paper. The results show the accuracy of the features. To extract more independent features for hand gestures recognition is the task of my future work.

References

1. Devi, M., Saharia, S., Bhattacharyya, D.K.: A dataset of single-hand gestures of Sattriya dance. In: Heritage Preservation 2018, pp. 293–310. Springer, Singapore
2. Devi, M., Saharia, S.: A two-level classification scheme for single-hand gestures of Sattriya dance. In: 2016 International Conference on Accessibility to Digital World (ICADW). IEEE, New York (2016)

Study of Various Methods for Tokenization

Abigail Rai and Samarjeet Borah

Abstract Tokenization is the mechanism of splitting or fragmenting the sentences and words to its possible smallest morpheme called as token. Morpheme is smallest possible word after which it cannot be broken further. As the tokenization is initial phase and as well very crucial phase of Part-Of-Speech (POS) tagging in Natural Language Processing (NLP). Tokenization could be sentence level and word level. This paper analyzes the possible tokenization methods that can be applied to tokenize the word efficiently.

Keywords Part-Of-Speech tagging (POS) · Tokenization · Natural Language Processing (NLP) · Morpheme · Token · etc.

1 Introduction

Natural Language Processing (NLP) is a blended area of research and application which involves primarily computer science and linguistics. It involves processing of the natural languages that a human understands and speaks to make it familiar with the machine, so human and machine can interact with each other efficiently. There are many working areas in NLP like Part-Of-Speech Tagging, Noun-Entity Recognition, Speech Recognition, and many more. In Natural Language Processing, Part-Of-Speech (POS) tagging is considered as the first step toward machine interaction. Tokenization is the initial step in Part-Of-Speech tagging. In tokenization, sentences are broken up into smaller meaningful units known as tokens. These also can be called as smallest individual units. In human languages, smallest units of words are words, punctuation mark, special characters, etc. Tokenization tokenizes by searching word

A. Rai (✉) · S. Borah
Department of Computer Application, SMIT, Sikkim Manipal Institute of Technology, Sikkim, India
e-mail: angellaraiz90@gmail.com

S. Borah
e-mail: samarjeetborah@gmail.com

J. K. Mandal et al. (eds.), *Applications of Internet of Things*, Lecture Notes in Networks and Systems 137, https://doi.org/10.1007/978-981-15-6198-6_18

boundaries in the sentences. Words boundaries are starting and end of the word. Another name of this process is segmentation.

This paper consists of the analysis of various research works completed by researchers. Initial section incorporates different approaches and algorithms used for tokenization in various researches, followed by the literature reviews and analysis of methods used.

2 Tokenization Approaches

2.1 Lucene Analyzer [1]

Lucene analyzers split the text into tokens. **Analyzers mainly consist of tokenizers and filters.** Different analyzers consist of different combinations of tokenizers and filters. Common Lucene analyzers are: stop analyzer, whitespace analyzer, standard analyzer, keyword analyzer, custom analyzer, and per field analyzer. To tokenize the given sentences into simpler tokens, the OpenNLP library provides three different classes:

Simple Tokenizer

- Simple Tokenizer creates an object of the respective class.
- Using the tokenize () method, sentences or text will be tokenized.
- Smaller tokens will be printed.

Whitespace Tokenizer

- Whitespace tokenizer initially starts by creating an object of its respective class.
- Tokenize () method will be used to tokenize the sentences.
- After partitioning the sentences into smaller meaningful chunks, it prints the tokens.

TokenizerME class

- Using the Tokenizer Model class, it loads the en-token.bin model.
- Instantiate the TokenizerME class.
- Tokenization of sentences can be done using tokenize () method of this class.

2.2 Byte Pair Encoding (BPE) [2]

In 2016, Byte Pair Encoding has been used to prepare sub-word dictionary. In 2019, Radfor et al. adopt Byte Pair Encoding to construct sub-word vector to build GPT-2.

GPT-2 [3] can predict the next word through trained corpus, and initially, it was tested with 40 GB of Internet text.

Algorithm 1: Algorithm for Tokenization using Byte Pair encoding

Step 1: Corpora Generation

Step 2: Define desirable sub-word vocabulary

Step 3: Split word to sequence of characters

3.1. Append suffix "</w>" with word frequency

Step 4: New sub word will be generated

Step 5: Loop step 4 until

5.1. If (sub word vocabulary size reached defined in step 2)

or

5.2. If (highest frequency is 1)

Step 6: End

2.3 *Word Piece [4]*

Word Piece is a word segmentation algorithm, and it is similar with Byte Pair Encoding. Schuster, and Nakajima introduced Word Piece by solving Japanese and Korea voice problem in 2012. Although, Word Piece is similar with Byte Pair Encoding, difference is the formation of a new sub-word by likelihood but not with the next highest frequency pair.

Algorithm 2: Algorithm for Tokenization using Word Piece

Step 1: Prepare training corpus

Step 2: Define a desired sub word vocabulary size

Step 3: Split word to sequence of characters

Step 4: Build Language Model based on step 3 data

Step 5: Select new word unit with increasing likelihood on training data most when added to the model

Step 6: Loop step 5 until

6.1. If (it reaches sub word vocabulary size defined in step 2) or

6.2. If (Likelihood reaches threshold value)

Step 7: End

2.4 Unigram Language Model [4]

For tokenization or sub-word segmentation Kudo. came up with unigram language model algorithm. Algorithm assumes that each sub-word occurrence is independent, and sub-word sequence will be the result of the product of sub-word occurrence probabilities. Unigram model works to build sub-word vocabulary.

Algorithm 3: Algorithm for tokenization with Unigram Model

Step 1: Generate large size training corpus

Step 2: Sub word vocabulary will be designed

Step 3: Optimize the probability of word occurrence by giving a word sequence

Step 4: Compute loss of each sub word

Step 5: Sort the symbol by loss and keep top X % of word (e.g. X can be 80). To avoid out-of- vocabulary, character level is recommended to be included as subset of sub word.

Step 6: If (sub word vocabulary size defined reached or changes in step 5 will not be applied)

6.1. Loop step 3

Step 7: Loop step 3 until

7.1. If (sub word vocabulary size defined is reached)

Or

7.2. If (no change in step 5)

Step 8: End

2.5 Critical Tokenization [5]

Critical tokenization uses the principle of maximum tokenization. Maximum tokenization has three sub-classes such as: "Forward Maximum Tokenization (FT), Backward Maximum Tokenization (BT), and Shortest Tokenization (ST)" [5]. Critical tokenization uses many mathematical concepts for tokenization process.

Some of the tokenization tools are:

- Word tokenization with python NLTK [6]
- Nipdotnet tokenizer [6]
- Mila tokenizer [6]
- NLTK word tokenizer [6]
- TextBlob word tokenizer [6]

- MBSP word tokenizer [6]
- Pattern word tokenizer [6].

3 Literature Review

This section contains review of conceptual literature of tokenization in NLP, and review of literature tried to analyzed most of the techniques used for tokenization for Indian and other languages.

Researchers have discussed ways to work with tokenization and pre-processing as important step for further work [7]. They analyzed and compared the working of different open-source tokenization tools. Concluded that Nlpdotnet tokenizer provided the best among seven tokenization tools compared, but still there is need to develop common tokenizer for all languages as existing tools are confined with limited languages.

How sentences can be tokenized using mathematical techniques has been described [5]. Initially, researcher introduces mathematical model which works with sentence generation and sentence segmentation. And he came up with distinctive work of developing a tokenization model, which works opposite of generation of sentence. Their findings and observations achieved so far have there still more work to be done.

By using tokenization and clustering [8], researchers successfully summarized the large volume of data. To accomplish their work text mining, they have implemented various processes like stop word removal and stemming.

Their research attempted to list out major categories of works under Natural Language Processing [9] and to understand the most used techniques.

Tokenizer works in two phases to complete tokenization, and for normalizing white spaces, pre-processing stage initially has been implemented and next for filtering tokens as post-processing stage [10]. Their work gives ideas to implement tokenization with different language depth.

In this [11] research, researcher concluded that tokenization complex or ambiguous will be disambiguated using Part-Of-Speech tagging.

In the research paper [12] by Okan Kolak et al., they worked on process that is end to end with the concept of channel with disturbance, generating true text by transforming into noisy output of an OCR system and concluded as their work provides much improvement. They are working in their research to make the same model efficient enough to work with other natural languages.

Their work presented a text or sentence normalizer to normalize Kannada text in machine translation system (MTS) [6]. The proposed text normalizer is tested on Enabling Minority Language Engineering (EMILLE) corpus, and nearly 45–57% of input text has been filtered during normalization itself.

For the development of the multi-word-tokenization (MWT) [13] as pre-processing-NLP, researchers have generalized the problem of single-word tokenization and multi-word.

The research work completed [14] presents tokenizer which tokenize sub-word and detokenize independent from any languages.

Researcher has developed NMT model [2], which can do encoding of exceptional and unfamiliar words as subsequent sub-words units which gives capability of translation of open vocabulary.

Completed research work [15] has successfully implemented the proposed work, i.e., sentence and token splitting using conditional random field (CRF). A tool with a linguistically adequate representation and a rich feature set can be employed for enhancement of their work.

Researchers have been contributed in the area of tokenization [16]. They have implemented low-level language-independent tokenizer which determines the word boundaries and tokenize words.

4 Analysis of the Tokenization Methods

Tokenization seems easy as it means to split words from sentences or words to get smallest meaningful token, but it has its complications. Complications related with tokenization differ from language to language:

- Hyphen and non-separating whitespace raise problems for tokenizing sentences or texts.
- When we separate by checking start and end of word boundaries, there might be a possibility of splitting a single word.
- In French language, they use apostrophe differently for reduced definite article as prior to word starts with vowel.
- There is more complication with the Chinese language as there are no word separators like space in most of other languages, and even has short words formed by two characters.

Researchers worked for tokenization of various languages achieved success in various complexity of tokenization processes like:

- They succeeded in splitting sentences into word and word into tokens using various techniques.
- Using mathematical models, they introduced sentence generation and sentence tokenization approach.
- Researchers successfully separated most concatenated words into distinct tokens, without losing inflection that appears on the word.

5 Features for Tokenization

- Beginning and end mark of the word—it works with already trained data's, so it is known start and end of the sentence or word and very well knows punctuation or character to mark with.
- Punctuation and spaces—it can be used to isolate the words with the help of spaces and punctuation. Punctuation cannot be discarded but what has to be done with it at the pre-processing time can be decided by the user.
- Features from different languages might vary accordingly with languages.

6 Conclusion

In this paper work, we tried to give a brief idea about the existing approaches that have been used to develop tokenizer. We have presented a survey on developments of different tokenization systems for Indian languages as well other languages. We found out from the survey that for various languages, rule-based, supervised, and unsupervised approaches have been used which have given good performance results. In each research work, the most task is to generate the most efficient tokenizer which can give the best performance for different languages.

References

1. Accessed on 6 Nov 2019. https://www.tutorialspoint.com/lucene/lucene_analysis.htm
2. Sennrich, R., et al.: Carried research on "Neural Machine Translation of Rare Words with Subword Units". arXiv: 1508.07909v5 [cs, CL] (10 June 2016)
3. Accessed on 6 Sept 2019. https://openai.com/blog/better-language-models/
4. Kudo, T.: Subword regularization: improving neural network translation models with multiple subword candidates. In: Proceedings of the 56th Annual Meeting of the Association for Computational Linguistics (Long Papers), pp. 66–75. Melbourne, Australia (July 15–20, 2018)
5. Guo, J., et al.: Critical Tokenization and its properties. Comput. Linguist. **23** (1997)
6. Prathibha, R.J., Padma, M.C.: Kannada text normalization in source analysis phase of machine translation system. Int. J. Eng. Technol. (IJET) **9**(3S) (2017). ISSN (Print): 2319–8613 ISSN (Online): 0975-4024. https://doi.org/10.21817/ijet/2017/v9i3/170903s088
7. Vijayarani, S., Janani, R.: Text mining: open source tokenization tools—an analysis. Adv. Comput. Intell. Int. J. (ACII), 3(1) (2016)
8. Joseph, J., Jeba, J.R.: Information extraction using tokenization and clustering methods. Int. J. Recent Technol. Eng. (IJRTE) **8**(4) (2019). ISSN: 2277–3878
9. Bulusu, A., Sucharita, V.: Research on machine learning techniques for POS tagging in NLP. Int. J. Recent Technol. Eng. (IJRTE), 8(1S4) (June 2019) ISSN: 2277–3878
10. Attia, M.A.: Arabic tokenization system. In: Proceedings of the 5th Workshop on Important Unresolved Matters, Prague, Czech Republic, pp. 65–72 (2007)
11. Barrett, Neil, Weber-Jahnke, Jens: Building a biomedical tokenizer using the token lattice design pattern and the adapted Viterbi algorithm. BMC Bioinform. **12**(Suppl 3), S1 (2011)
12. Kolak, O.: A generative probabilistic OCR model for NLP applications. In: Proceedings of HLT-NAACL Main Papers, p. 55. Edmonton (May–June 2003)

13. Schütze, H., Padó, S.: Multi-word tokenization for natural language processing. **14** (2013)
14. Kudo, T., Richardson, J.: SentencePiece: a simple and language independent subword tokenizer and detokenizer for neural text processing. In: Proceedings Conference on EM in NLP (System Demonstrations), pp. 66–71. Brussels, Belgium (October 31–November 4, 2018)
15. Tomanek, K., et al. Sentence and token splitting based on conditional random fields. CPACL (2007)
16. Megerdoomian, K., Zajac, R.: Processing Persian Text: Tokenization in the Shiraz Project. Memoranda in Computer and Cognitive Science MCCS-00-322 (April 2000)

A Categorical Study on Cache Replacement Policies for Hierarchical Cache Memory

Purnendu Das and Bishwa Ranjan Roy

Abstract Cache memory plays an important role in the in-memory computation in memory-intensive applications. Hierarchical cache design is used to increase the capacity of cache to handle large working set. The last level cache (LLC) does not strictly follow the temporal locality of program, so it becomes challenging to identify the blocks that will not be reused (dead block). In this paper, we have performed a detail survey on different techniques to detect the dead blocks early in the cache memory and improve the hit rate of cache replacement algorithm. Belady's optimal solution detects the dead block by analyzing the future of blocks, which is completely un-realistic. Many researches have been done to detect dead block practically by observing the previous access pattern. Many algorithms are proposed to improve the performance of traditional replacement policies by considering different additional information. Most of the algorithm aims to reduce the miss count by retaining the blocks that will be reused before eviction (live blocks). Recent study observes that the cost of all the cache miss are not uniform in nature. So, some researchers have distinguished between high-cost block and low-cost block. The overall cost can be reduced by retaining the high-cost block in memory, with little higher miss count. It is observed that by managing cache miss un-coordinately among the different levels of cache memory, it is not possible to obtain maximum utilization of memory. Many adaptive algorithms have been proposed to maintain balance between the over-utilized blocks and underutilized blocks by the displacement of blocks. In this survey, we have categorized the practically implemented techniques into different classes based on their basic principle of cache replacement.

Keywords Hierarchical cache · Replacement policy · Last level cache · Dead block

---

P. Das · B. R. Roy (✉)
Department of Computer Science, Assam University Silchar, Silchar, Assam, India
e-mail: brroy88@gmail.com

P. Das
e-mail: purnen1982@gmail.com

J. K. Mandal et al. (eds.), *Applications of Internet of Things*, Lecture Notes in Networks and Systems 137, https://doi.org/10.1007/978-981-15-6198-6_19

1 Introduction

The importance of in-memory computing is increasing day-by-day to handle big-data application. Data-intensive application works on large amount of data with larger size of working set resulting the demand of increase in the size of frequently accessed data blocks. Due to the increasing performance gap between the modern processor and the memory [1], it becomes crucial to design cache hierarchy to increase the capacity of on-chip memory. Each core in CMP system maintains a private section of cache memory and shares another large section of cache memory with all other cores. The private cache memory of each core is called L1 cache, and the shared cache memory is known as L2 cache or last-level cache (LLC) as shown in Fig. 1. Cache memory is used to maintain frequently accessed data closer to the processor to reduce latency. Though the last-level cache may suffer a high access latency, it is negligible compared to the access latency to main memory. Pipelining and parallel processing are implemented based on L1 cache, so levels below L1 are less critical to handle.

The increase in the capacity of cache will not improve the cache hit until a better replacement policy is used to allocate space for blocks on demand. A cache block that will be accessed again before eviction is called live block, and a block that will not be accessed anymore by before eviction is called dead block [2–4]. The main objective of replacement algorithm is to retain live blocks and remove dead block immediately from the cache. Many researches have been done to predict dead block at the earliest but failed to meet the performance of Belady's optimal solution [5]. Efficient cache replacement policy can increase the hit rate by retaining on demand blocks. In some situation, cost of access became crucial to manage compared to hit rate. Some work has been done to reduce the overall cost by allowing the higher miss rate.

Fig. 1 Architecture of chip-multiprocessor

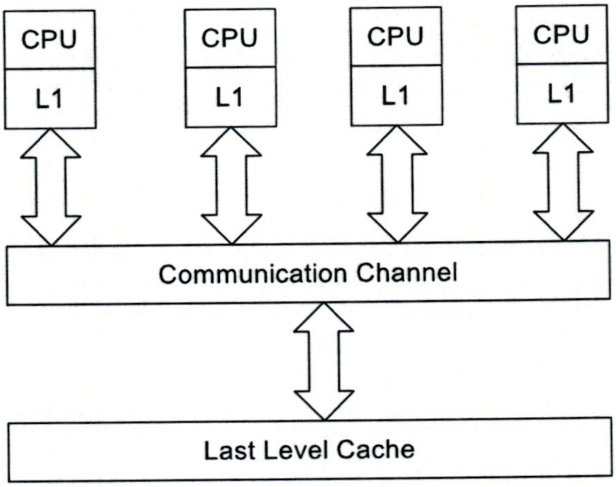

In this paper, we have discussed the basic principle of cache replacement and their limitations. We have classified the available replacement policies and analyzed the mechanism and performance in the respective class.

2 Basic Replacement Policy

The goal of a replacement policy is to evict a block that will not be accessed in near future. Any replacement policy follows these basic steps: (1) Victim selection: selection of the block to be evicted to accommodate recently access block. (2) Block insertion: where to place the recently accessed block. (3) Block promotion: on what basis the priority of a block will be increased.

The Belady's optimal replacement OPT algorithm [5] is considered as benchmark algorithm as it gives minimum cache miss. Belady's algorithm uses the future knowledge to select the victim block. So, it remains theoretical, and it is not feasible to implement. Mettension proposed an OPT to compute optimal miss count of a trace. LRU replacement policy is considered as the most efficient and is widely accepted replacement policy. In LRU policy, it is predicted that the recently accessed block will be referred soon. So, the newly requested block is placed at the most recently used (MRU) position. In direct, the most frequently accessed block remains in close proximity to the MRU position. But in real- time application, it is not always true. A block may be accessed frequently for very short period of time, then after it may not be accessed again. There may be single-time used block also. In both cases, a block occupies space unnecessarily until it reaches the LRU position. To avoid drawbacks of LRU, other primitive algorithms are proposed, namely MRU replacement policy and random replacement [6]. MRU replacement policy selects the most recently accessed block as victim with the prediction that the recently accessed block will not be accessed soon. It performs better for a sequence of single-access blocks. The random replacement policy selects the victim block randomly without considering recent history. In some cases, random replacement performs better.

3 Tradeoff of Traditional Replacement Policy

It is required to increase the cache associativity to avoid conflict miss, and on the other hand, the necessity of efficient replacement algorithm increases with the increase in the associativity. Recent studies observed large performance gap between OPT and LRU in case of highly cache associativity. OPT uses perfect knowledge to select victim block, whereas practical replacement policy predicts the reuse of the block based on available past knowledge. In case of memory-intensive applications, LRU fails to reduce cache miss as the size of working set exceeds the capacity of cache [7]. In multilevel cache, levels below L1 do not strictly follow the program locality. So LRU fails to detect dead block for eviction. The traditional LRU replacement policy uses only recency of access while LFU replacement policy uses frequency of access to select the victim. Absence of other information causes problem like thrashing in LRU and aging in LFU.

4 Improved Replacement Policies

Initially, some researchers have attempted to improve the traditional replacement policies by making minor changes to the replacement policy, insertion policy, and promotion policies. Belady's algorithm gives the best cache miss reduction, but it is purely theoretical as it require future knowledge to select the victim. In this paper [8], the author attempted to generate the future knowledge from the past to evict dead block. To do this, author first reconstructed the Belady's optimal solution from the past cache reference history and then analyzed this reconstructed information to predict the victim block. The authors have used set dueling method [9] to design this algorithm.

Another author has proposed a replacement policy based on tag distance correlation among cache lines in cache set [10]. This replacement policy finds the victim block by considering the LRU behavior bit instead of replacing LRU lines straight forward. So, LRU lines get a chance to remain in the cache for some more time. MIP does not follow the LRU line to evict a block directly, and it combines the LRU ordering with promotion policy, to achieve adaptive insertion mechanism. This method uses set dueling [9] and dynamic set sampling to improve cache hit rate and to reduce hardware overhead. The sets in each block are split into groups of size *n*, where *n* is power of 2. Each group is again divided into two different types of sets. The purpose of evaluation set is to select more appropriate insertion position so that the optimal hit rate can be achieved. The international policy of convolution set is estimated based on the performance of evaluation set.

A new family of LIFO is proposed to overcome the limitation of LRU [11]. In some situation, the block placed in MRU position itself becomes dead and occupies a memory block for a large interval. They classified this large class into (1) dead block prediction LIFO, (2) probabilistic escape LIFO, and (3) probabilistic counter LIFO. Pseudo-LIFO algorithm prefers the eviction of block residing at the top of the fill stack. Different member uses additional information to select the better victim block. In this paper [12], the authors have proposed a hybrid model by exploiting the advantages of peLIFO and LRU.

5 Reuse Distance Based Algorithm

The reused based replacement policy attempts to identify and exploit reuse locality effectively. To achieve these objectives, replacement algorithm tries to exploit temporal locality. Whenever a block is in the private cache, it is assumed that the

Fig. 2 Replacement order based on reference

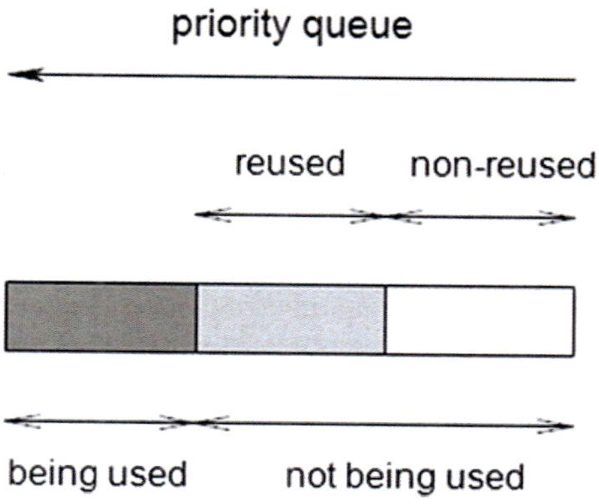

temporal locality is well exploited, so the block must be retained in the LLC. Among the blocks not present in the private L1 cache, replacement algorithm will select victim one based on the reuse order (Fig. 2).

Albericio [13] have proposed two simple replacement algorithm exploiting reuse locality to establish the replacement order: (1) least recently reused (LRR) and (2) not recently reused (NRR). Hardware cost is same as NRU while LRR require an additional bit per line. These algorithms try to retain blocks in the LLC that is present in the private L1 caches expecting the future re-reference. The recently used is placed into the MRU position, so it will be the last candidate to be replaced. Initially, these algorithms search for the blocks that are not residing in the private caches and also *non-reused* are selected as victim randomly. If there are no such block, algorithms search for a block that is not residing in the private caches but *reused* is selected as victim, and lastly, a block which is *being used* in the private caches is selected for eviction.

LRFU replacement policy is introduced by KIM [14] which is a combination of least frequently used (LFU) and least recently used (LRU). The traditional LRU replacement policy analyzes the recent pattern of block access to select the victim. However, the least frequently used (LFU) policy keeps track of older history of block access. LRU can be implemented by recording the timestamp of block access. The smallest the time stamp, oldest in access. LFU maintains a counter for each block. Upon every reference to a block, the counter of that block will be incremented. A block with smallest counter value means it is the least frequently used block. LRFU extracts the advantages of both the policies by calculating a weight factor combined recency and frequency (CRF). CRF is proportional to the recency of reference.

Denning et al. distinguish optimal replacement algorithm based on whether the future information is unrealizable or realizable [15]. His optimal algorithm makes the best possible replacement decision based on a statistical model that is used to understand the future program behavior accurately. Other prefetching algorithm tries to minimize the overall misses by prefetching the blocks likely to be used in future. The algorithm tries to load the blocks on demand instead of performing direct prefetching.

Belady's MIN evicts the block that is reused furthest in the future while demand MIN tries to evict the block prefetched furthest in the future [16].

6 Counter-Based Algorithm

A counter-based algorithm [17] is proposed by Mazen and Yan to evict dead block efficiently. This policy determined the reuse distance (the interval between the first access to the next access) to determine the victim block. A block with maximum reuse distance is selected as the victim. Two different techniques are followed to implement this policy: (i) access interval predictor (AIP) and (ii) live-time predictor (LVP). In access interval predictor, the counter is incremented on every access to the same set during an access interval of block. Same set is considered to reduce the counter size as well as to avoid the access behavior of other block. The maximum value of the counter is assigned to the threshold. Live-time predictor keeps record of interval between the first access and the last access of block during its generation time. The period after last access is considered as dead time. The longest interval is taken as the threshold value.

Counter-based cache replacement is implemented by augmenting each block of cache with an event counter. A cache hit on the same set increases the counter value. Once the counter value exceeded the threshold value, the respective cache block becomes evictable. So, the blocks that have higher chance to become dead are evicted early from the cache creating more space for useful line. In this paper [18], the authors have observed that number of expected hits of a cache block have a strong relationship with the reciprocal of the reuse distance of that block. They have utilized this information to design an efficient algorithm to select the victim block with low cost. The algorithm is implemented based on counter to reduce the hardware cost. On each hit, counter is incremented to keep track of reuse distance. The victim block is selected by looking into the counter values.

H. Liu has proposed a new class of dead block prediction based on burst of access to cash blocks instead of individual references [19]. The cache burst is the period of a block spent in the MRU position. With this efficient dead block, the cache efficiency is improved by two ways, (1) cache optimizing and (2) bypassing and prefetching. The performance of the dead block predictor improves with the use of prefetching. In this paper, [18] by predicting the never re-accessed blocks, L2 cache is bypassed by placing this block directly in L1.

7 Cost-Based Replacement Algorithms

Most of the cache replacement algorithms have focused to reduce miss count. But in practical situation, cost of all the misses are not equal [20, 21]. The modern super scalar processor has the ability to hide private cache miss penalty by exploiting instruction-level-parallelism, but unfortunately, it is quite impossible to hide the large shared LLC cache miss penalty.

A. Jain has proposed an algorithm to handle the cache miss of different cost [22]. This algorithm considered two factors: locality and cost sensitivity, and simultaneously, they named it LACS. Cost is estimated from the no. of instruction managed by the processor to issue on that block. A block with low cost and poor locality is selected as victim. The LACS algorithm distinguishes the cache blocks based the cost associated with it. The algorithms try to retain maximum number of high-cost block in the cache. A locality algorithm is applied to revert the cost of blocks from high to low if it is not accessed. The block with minimum cost is always selected as the victim block. In NUMA LLC architecture in multiprocessors, Jeong and Dubois [23, 24] observe that cost associated with a remote block access is considerably higher than a neighbor block in terms of latency, bandwidth, and energy consumption also. Their proposed cost-sensitive replacement algorithm improves the overall cost of miss latency compared to OPT even though miss count is compromised.

Later Jeong [25] has taken the advantages of cost associated with load (high cost) and store (low cost) to propose a cost-sensitive algorithm to predict whether the immediate access will be a load or store instruction. The algorithm assumes the uniform cost for all the load instruction and manages to reduce miss cost by avoiding miss associated with load. Srinivasan et al. [26] designed a cache architecture to preserve the critical block or to perform prefetching of critical blocks. Critical blocks are identified by analyzing load chain and the ability of processor to execute independent instruction forwarded by the load instruction. Young have proposed greedy dual algorithm for cache management in network environment. In this algorithm, the author assumes uniform sized document associated with different cost. The algorithm manages to keep document with higher cost in the cache memory while evicting the document with lower cost. Authors have considered another factor weighted frequency based time to improve the performance of GDA [27].

To minimize energy consumption, the authors have proposed hybrid cache architecture composed of non-volatile memory (NVM) and DRAM. The MALRU (Miss-latency Aware LRU) [28] cache replacement algorithm tries to retain NVM block (high latency) in memory and preferentially selects victim from the DRAM block (low latency). Simultaneously MALRU keep on updating the reserve section of DRAM blocks to improve the performance.

8 Adaptive Replacement Policy

Tian et al. [29] have analyzed the number of occurrence (frequency) and the time of occurrence (recency) from cache history to predict most suitable victim block. They have designed an analytical model effectiveness-based replacement (EBR) policy which uses these data to form ranking of blocks within each set and replace the blocks with the lowest rank. EBR maintains higher weightage of recency compared to frequency. It scales the recency stack into different levels to generate independent subgroup. Frequency is used to set ranking inside each set. Set dueling [30] method is used to provide dynamic behavior to EBR by allowing dynamic generation of subgroup in recency stack.

Quereshi et al. have observed that a simple change to the insertion policy of the replacement algorithm can significantly decrease the number of misses for memory-intensive applications [9]. They have proposed three algorithms and analyzed the performance. Initially, they proposed LRU insertion policy (LIP) in which the incoming block is placed in the LRU position instead of MRU position. LIP is thrashing resistant algorithm with minimal cost, and its performance is close enough to optimal hit rate. They proposed bimodal insertion policy (BIP) by enhancing the LIP policy adaptable to changes in the working set without compromising the thrashing protection of LIP. Finally, they proposed a dynamic insertion policy (DIP) to select best-suited insertion policy out of BIP and traditional LRU that can reach minimal misses.

In global cache memory management [31], static information of the system is combined with the dynamic information to detect the dead blocks. If at some levels of cache, the blocks are assumed to be dead, and all the blocks of that level can be evicted immediately. In hierarchical cache architecture, managing a single level cannot achieve the maximum utilization of the total space, so it is important to take the replacement decision by coordinating all the levels of cache.

Re-reference interval algorithm does not consider priority of all the blocks in the set to control the priority queue on a cache miss. Adaptive demolition policy [32] estimates a subtraction value which will be subtracted from all the blocks on each cache miss. ADP considered the half of the average of the priority value to be minimized or unchanged. The authors have proposed a reference-table-based LRU algorithm to evict the dead blocks [33]. This algorithm is adaptive to the workload and changes the cache access based on set access pattern. D. Rolan proposed set balancing cache replacement algorithm in which the consequences of non-uniform distribution of memory in the cache set [34]. It is observed that some working set is bigger than the available cache set while other working set is small enough to fit into that cache set. In such case, the cache set may remain underutilized. The SDC aims to balance the load of cache set by associating other sets.

9 Conclusion

The purpose of cache memory is to keep frequently accessed memory blocks close to the processor. The program locality may grow beyond the capacity of cache in big-data application. Efficient algorithm can preserve the frequently acceded blocks in the cache while evicting the deadlocks immediately. In this paper, we have classified the replacement algorithm based on their approaches. Most of the algorithm focused on the reduction of miss rate. Some authors have achieved significant improvement in the performance by bringing small changes in traditional algorithms. Probabilistic approach is also used to select the victim block more accurately. Many researchers attempted to reduce the cost of access latency without limiting cache miss. They observed that a few number of high-cost block miss can decline the performance significantly. To achieve the maximum utilization of total cache capacity, adaptive approaches are followed to select the best -suited algorithm for specific cache access pattern. More effective replacement algorithm is still required to achieve the performance equivalent to benchmark algorithm OPT.

References

1. Wulf, W.A., McKee, S.A.: Hitting the memory wall: implications of the obvious. Comput. Archit. News **23**, 20–24 (1995)
2. Lai, A.-C., Fide, C., Falsafi, B.: Dead-block prediction & dead-block correlating prefetchers. ACMSIGARCH Comput. Archit. News **29**(2), 144–154 (2001)
3. Liu, H., Ferdman, M., Huh, J., Burger, D.: Cache bursts: a new approach for eliminating deadblocks and increasing cache efficiency. In: 41st IEEE/ACM International Symposium on Micro-architecture, 2008, pp. 222–233
4. Das, P.: Role of cache replacement policies in high performance computing systems: a survey. Commun. Comput. Inform. Sci. 400–410 (2019)
5. Belady, L.: A study of replacement algorithms for a virtual-storage computer. IBM Syst.J. (1966)
6. Das, S., Polavarapu, N., Halwe, P.D., Kapoor, H.K.: Random-LRU: a re-placement policy for chip multiprocessors. In: Proceedings of the International Symposium on VLSI Design and Test (VDAT) (July 2013)
7. Roy, B., Das, P.: SplitWays: an efficient replacement policy for larger sized cache memory. Int. J. Eng. Adv. Technol. (IJEAT), **9**(1), 4230–4234 (October 2019)
8. Jain, A., Lin, C.: Back to the future: leveraging Belady's algorithm for improved cache replacement. ACM/IEEE 43rd Annual International Symposium on Computer Architecture (ISCA), Seoul, pp. 78–89 (2016)
9. Qureshi, M.K., Jaleel, A., Patt, Y.N., Steely, S.C., Emer, J.: Adaptive insertion policies for high performance caching. In: International Symposium on Computer Architecture (ISCA), pp. 381–391 (2007)
10. Do, C.T., Choi, H.-J., Kim, J.M., Kim, C.H.: A new cache replacement algorithm for last-level caches by exploiting tag-distance correlation of cache lines. Microprocess. Microsyst. **39**(4–5), 286–295 (2015)
11. Chaudhuri, M.: Pseudo-LIFO: the foundation of a new family of replacement policies for last-level caches. In: Proceedings of the 42nd Annual IEEE/ACM International Symposium on Microarchitecture (MICRO 42). ACM, New York, NY, USA, 2009, pp. 401–412
12. Rodríguez-Rodríguez, R., Castro, F., Chaver, D., Pinuel, L., Tirado, F.: Reducing Writes in Phase-Change Memory Environments by Using Efficient Cache Replacement Policies, pp. 93–96. Design, Automation & Test in Europe Conference & Exhibition (DATE), Grenoble, France (2013)
13. Albericio, J., Ibáñez, P., Viñals, V., Llabería, J.M.: Exploiting reuse locality on inclusive shared last-level caches. ACM Trans. Archit. Code Optim. **9**(4), 19 (January 2013). Article 38
14. Lee, D., Choi, J., Kim, J., Noh, S., Min, S., Cho, Y., Kim, C.: LRFU: A spectrum of policies that subsumes the least recently used and least frequently used policies. IEEE Trans. Comput. **50**(12) (2001)
15. Denning, P.J.: Thrashing: its causes and prevention. In: Proceedings of the December 9–11, 1968 Fall Joint Computer Conference Part I, pp. 915–922 (1968)
16. Jain, A., Lin, C.: Rethinking Belady's Algorithm to Accommodate Prefetching, pp. 110–123. ACM/IEEE 45th Annual International Symposium on Computer Architecture (ISCA), Los Angeles, CA (2018)

17. Kharbutli, M., Solihin, Y.: Counter-based cache replacement and bypassing algorithms. IEEE Trans. Comput. (2008)
18. Vakil-Ghahani, A., Mahdizadeh-Shahri, S., Lotfi-Namin, M., Bakhshalipour, M. Lotfi-Kamran, P., Sarbazi-Azad , H.: Cache Replacement Policy Based on Expected Hit Count. IEEE Computer Architecture Letters **17**(1), 64–67 (2018)
19. Liu, H., Ferdman, M., Huh, J., Burger, D.: Cache bursts: A new approach for eliminating dead blocks and increasing cache efficiency. 41st IEEE/ACM International Symposium on Microarchitecture, Lake Como, 2008, pp. 222–233
20. Qureshi, M., Lynch, D., Mutlu, O., Patt, Y.: A case for MLP-aware cache replacement. In: Proceedings of 33rd Annual International Symposium Computer Architecture, pp. 167–178 (2006)
21. Kharbutli, M., Sheikh, R.: LACS: a locality-aware cost-sensitive cache replacement algorithm. IEEE Trans. Comput. **63**(8), 1975–1987 (2014)
22. Sheikh, R., Kharbutli, M.: Improving cache performance by combining cost-sensitivity and locality principles in cache replacement algorithms. In: IEEE International Conference on Computer Design, Amsterdam, pp. 76–83 (2010)
23. Jeong, J., Dubois, M.: Cost-sensitive cache replacement algorithms. In: Proceedings of 9th Interational Symposium High-Perform. Computer Architecture, pp. 327–337 (2003)
24. Jeong, J., Dubois, M.: Cache replacement algorithms with nonuniform miss costs. IEEE Trans. Comput. **55**(4), 353–365 (2006)
25. Jeong, J., Stenstrom, P., Dubois, M.: Simple penalty-sensitive cache replacement policies. J. Instruct.-Level Parallel **10** (2008)
26. Srinivasan, S., Dz-Ching Ju, R., Lebeck, A., Wilkerson, C.: Locality vs. criticality. In: Proceedings of 28th Annual International Symposium Computer Architecture, pp. 132–143 (2001)
27. Ma, T., Qu, J., Shen, W., Tian, Y., Al-Dhelaan, A., Al-Rodhaan, M.: Weighted greedy dual size frequency based caching replacement algorithm. In IEEE Access, vol. 6, pp. 7214–7223 (2018)
28. Chen, D., Jin, H., Liao, X., Liu, H., Guo, R., Liu, D.: MALRU: Miss-penalty aware LRU-based cache replacement for hybrid memory systems. Design, Automation & Test in Europe Conference & Exhibition (DATE), Lausanne, pp. 1086–1091 (2017)
29. Tian, G., Liebelt, M.: An effectiveness-based adaptive cache replacement policy. Microprocess. Microsyst. **38**(1), 98–111 (2014)
30. Qureshi, M.K., Jaleel, A., Patt, Y.N., Steely, S.C., Emer, J.: Adaptive insertion policies for high performance caching. In: Proceedings of the 34th Annual International Symposium on Computer Architecture, San Diego, California, USA (2007), 09–13 June 2007
31. Manivannan, M., Pericás, M., Papaefstathiou, V., Stenström, P.: Global dead-block management for task-parallel programs. ACM Trans. Archit. Code Optim. **15**(3), 25 (2018), Article 33
32. Tada, J., Sato, M., Egawa, R.: An adaptive demotion policy for high-associativity caches. In: Proceedings of the 8th International Symposium on Highly Efficient Accelerators and Reconfigurable Technologies (HEART 2017). ACM, New York, NY, USA (2017), Article 4, 6
33. Reishi Kumaar, T., Sharma, A., Bhaskar, M.: Reference table based cache design using LRU replacement algorithm for Last Level Cache. In: IEEE Region 10 Conference (TENCON), Singapore, pp. 2219–2223 (2016)
34. Rolán, D., Fraguela, B.B., Doallo, R.: Adaptive line placement with the set balancing cache. In: 42nd Annual IEEE/ACM International Symposium on Microarchitecture (MICRO), New York, NY, 2009, pp. 529–540

Dr. Purnendu Das is an assistant professor of the Department of Computer Science, Assam University, Silchar. He has pursued Ph.D. Degree from Tripura University. He has published research papers in many reputed journals.

Bishwa Ranjan Roy is an assistant professor of the Department of Computer Science, Assam University, Silchar. He has done M.Tech degree at NIT Silchar. He has many publications in many reputed journals.

Side-Channel Attack in Internet of Things: A Survey

Mampi Devi and Abhishek Majumder

Abstract To ensure security for data exchange is a challenges task in Internet of Things (IoT). Thus, research on side-channel attack is a major issue in this domain. Side-channel attack is based on side-channel information. This attack is of either ciphertext only attack or plaintext only attack or chosen plaintext attack. Moreover, since this attack is cheap to perform, it requires little computing power and is relatively easy to perform. So, this attack is growing day by day. Therefore, security is not an easy task to establish in a given system. The motive behind this paper is to present a comprehensive survey on different types of IoT attack with special focus on side-channel attack. In addition, a list of research issues and open challenges are also highlighted in the paper.

Keywords Internet of things · Side-channel attack · Internet security · Cryptography

1 Introduction

Recently, the world is more connected through the electronic devices specially known as Internet of things (IoT) technology. Ashton [1] is the pioneer of the term IoT. Internet of things is instance technological changes which represent the future of computing and communications. To develop this technology is a dynamic invention, which is spread from wireless sensors network field to the nanotechnology-based architecture [2–4]. Nowadays, this IoT has potential applications spreading from smart city, control actuation and maintenance of complex systems in industry field to health transport. In simple, we can say that IoT becomes a important part of our lives.

M. Devi (✉) · A. Majumder
Department of Computer Science and Engineering, Tripura University, Suryamaninagar, Agartala, Tripura 799022, India
e-mail: drammpidevi@gmail.com

A. Majumder
e-mail: abhi2012@gmail.com

J. K. Mandal et al. (eds.), *Applications of Internet of Things*, Lecture Notes in Networks and Systems 137, https://doi.org/10.1007/978-981-15-6198-6_20

Internet of things is unquestionably an emerging trend of research area. Though it is a very young field, there are huge amount of research and development that have taken place on IoT. Thus, various vulnerabilities have been shown throughout the use of IoT and so, this current technology is in danger situation. And, due to this reason, attacks on IoT were invented. Definitely, security of data exchange is a challenging task for Internet of things.

Side-channel attack (SCA) is one of the important attacks during data exchange in IoT because this attack can be easily performed and required less power consumption. The first official information related to SCA attack dates back to 1965 [5]. The side-channel attacks are the attacks which are based on side-channel information. This attack is of either ciphertext only attack or plaintext only attack or chosen plaintext attack. Before knowing the side-channel attack (SCA), it is necessary to have knowledge about side-channel information [1]. This information is not related to any encryption and decryption devices. The side-channel information perceived as a unit in which input is plaintext and output is ciphertext and vice versa. Different types of side-channel attacks are (i) timing attack, (ii) power analysis attack which has also subtypes such as simple power analysis (SPA) attack, deferential power analysis (DPA) attack, and co-relation power analysis (CPA) attack, and (iii) fault attack [2]. This type of attack instead of attacking the mathematical properties of the algorithm takes the advantages of physical phenomena that occur when cryptography algorithm implemented in hardware. There are different types of cryptography algorithm that is available in the literature. Some of them are Advanced Encryption Standard (AES), Data Encryption Standard (DES), and Elliptic Curve Cryptosystem (ECC).

The main aspects of this survey is to provide information about different attacks in IoT domain with special focus on side-channel attacks. In addition, how the attacks are performed and list out the issues to provide ideas for future scope in this domain.

The paper is organized as follows. Section 2 discusses about taxonomy of attacks (classification of attack) on IoT. Section 3 gives an overview of cryptanalysis, i.e., different types of cryptography algorithm which are the prerequisite knowledge of side-channel attack. In Sect. 4, a comprehensive study on side-channel attacks has been presented. Section 5 highlighted the open issues to do further research in this domain and finally, the paper concludes with Sect. 6.

2 IoT Attacks Taxonomy

With the increasing use of IoT devices, the security issues of IoT also increases. In general, the IoT attacks are broadly classified into five classes, viz. physical attacks, side-channel attacks, cryptanalysis attacks, software attacks, and network attacks. Apart from this broad classification, several other subclassifications are provided under this classification as shown in Fig. 1.

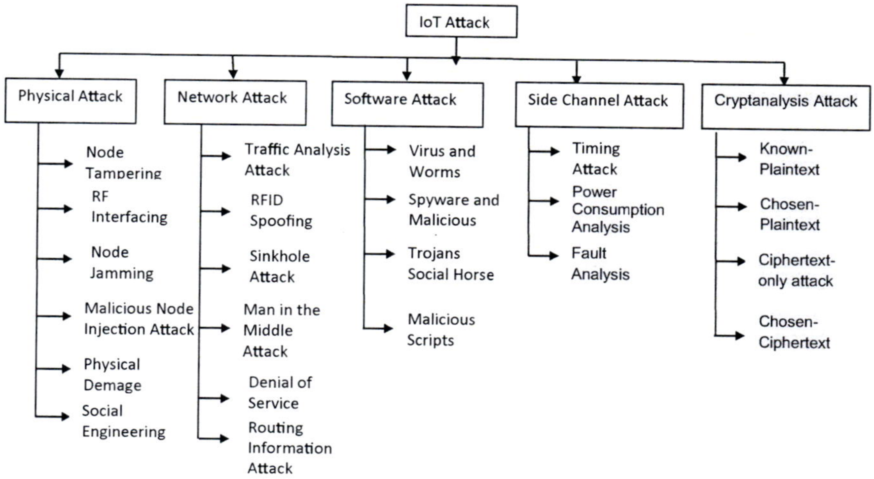

Fig. 1 Side-channel attacks classification: a taxonomy

2.1 Physical Attacks

These types of attacks are related to the hardware component. It is a hard attack due to its expensive material. Example of this attacks are depackaging chip, microprobing, etc. The physical attacks are of the following types [6]

(i) **Node Tempering Attacks**: In this attack, the attacker physically attacks the trusted node and can obtain important information.
(ii) **Rf Interference Attacks**: These types of attacks perform denial-of-service attack by sending noise signal over radio frequency.
(iii) **Node Jamming Attacks**: These types of attacks are performed in wireless sensor network where the attacker attacks the wireless communication by using jammer. These types of attacks also cause the deniel-of-service attack.
(iv) **Malicious Node Injection Attacks**: In this type of attack, the attacker injects a malicious node between two or more different nodes. Thus, the malicious node modifies the data and passes the wrong information to the other node.
(v) **Physical Damage**: Here, the attacker physically harms the component and resulting deniel-of- service attack.
(vi) **Social Engineering Attacks**: This type of attacks occurs when attacker interacts physically and manipulates users of an IoT system.

2.2 Network Attacks

Because of the broadcast nature of the transmission, wireless communication systems are vulnerable to network security attacks. Basically, attacks are classified as active and passive attacks. The passive attacks are the attacks where the physical, electrical effects of the functionality of the devices are used. The network attacks include monitor and eavesdropping, traffic analysis, camouflage adversaries, etc. On the other way, in case of active attack, attacker has to reach the internal circuitry of the cryptographic devices. Some of these types of attacks are denial-of- service attacks, node subversion, node malfunction, node capture, node outage, message corruption, false node, and routing attacks.

2.3 Software Attacks

Software attack occurs when malware is installed in networks program. This malicious software includes a virus, corrupted data and mainly involves injection of malicious code into the system. A software attack can also launch a DDoS attack. For example, jamming which is the largest threat of Internet of things. Here, small network exists which consists of small number of nodes including less amount of energy consumption and resources.

2.4 Side-Channel Attacks

In cryptography, a side-channel attack is based on information gain from on physical implementation of crypto system. Examples of side-channel attacks are timing attacks, power consumption analysis attacks, fault analysis attacks, electromagnetic attacks, and environmental attacks. Side-channel attacks are briefly described in Sect. 3. Some of the common types of side-channel attack are as followings:

(i) **Timing Attacks**: These attacks are calculated by measuring the time taken for unit operation perform.
(ii) **Power Consumption Analysis Attacks**: These attacks depend on power consumption analysis during encryption operation perform. These types of attacks are subdivided into simple power analysis attack and co-relation power analysis attack.
(iii) **Fault Analysis Attacks**: Fault analysis attacks are the recent and more powerful cryptanalysis attack to perform some faulty operations, with the expectation that the results of the fault operation will leak information about the secret keys involved.

2.5 Cryptanalysis Attacks

Cryptanalysis attack is the attack where the attacker known the encryption key by using either plaintext or ciphertext. According to the state of the art [6], there are different types of cryptoanalysis attacks available such as (i) known-plaintext attack, (ii) chosen-plaintext attack, (iii) ciphertext-only attack, (iv) chosen-ciphertext attack, and (v) chosen-key attack A cryptosystem is a system where pair of communicating occurs with the assurance of security. The cryptosystem consists of five touples (P, C, K, E, D), where P is all the set of plaintext ($_e$), i.e., the message between the two pairs. C is all the set of ciphertext ($C_k(_e+K)$) and K is the secret key shared by two pairs. E is the encryption key, and D is the decryption key. The security of a system fully depends on the secrecy of the secret key. Analyzing a cryptosystem to find a weakness that would leak the secret key is called cryptanalysis [7]. Some of the most popular cryptographic algorithms existing in literature are namely Advanced Encryption Standard (AES), Advanced Encryption Standard (DES), and Elliptic Curve Cryptography (ECC). These cryptographic algorithms are shortly described as follows:

Advanced Encryption Standard: Advanced Encryption Standard (AES) is a block cipher [6] cryptographic algorithm. As every block cipher the AES algorithm takes two inputs: a block of data with length n bits and a key with length k bits. There is a deterministic relationship between input and output. For AES, the input data and key can be 128-bit, 192-bit, or 256-bit long independent of each other. The Federal Information Processing Standards (FIPS) are standards published by the US government. These are made for public use, and especially non-military US government agencies. FIPS-197 [8] was announced in 2001. It specifies how the US government should use AES. FIPS-197 specifies that the data blocks used in AES always are be 128 bits, but key length could be 128-bit, 192-bit, or 256-bit.

Data Encryption Standard (DES): DES algorithm is a symmetric-key block cipher cryptographic algorithm where same keys are used in both encryption and decryption operation. This DES algorithm first published in the National Institute of Standards and Technology (NIST). This algorithm has 16 round Feistel structure and 64-bit block size. Out of 64-bit, 56 bits are effective key length, The 8-bit keys are not used during encryption performance(function as check bits only) [9].

Elliptic Curve Cryptography (ECC): ECC is the very latest encryption algorithm which was discovered in 1985 by Victor Miller [7]. It is the most secured encryption algorithm compared to other existent RSA and DSA algorithm. Compared to the other algorithms, the 256-bit of ECC is equal to 3072-bit RSA key. Since it is a very short key, also used less computational power and fast and secure connection, it is the most ideal algorithm for smart phone and tablet too [10]. Though ECC is a very small but due to its proper security it gain its popularity day by day. However, ECC certificates key creation method is different from previous algorithms. To see the popularity of elliptic curves, we can likely say that it is next generation of cryptographic algorithms, and we are beginner of their use now. Due to the ability of ECC algorithm to run efficiently on different hardware's (from 8-bit smart cards

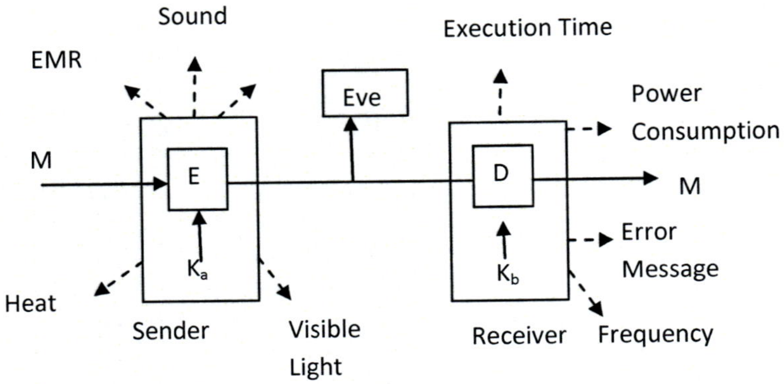

Fig. 2 Block diagram of side-channel attack

to high end computers) makes it suitable for IoT applications. This makes it suitable for IoT.

3 Side-Channel Attack

In side-channel attack (SCA), the eavesdropper can be able to monitor the power consumed during smart card operation, electromagnetic radiation during performance of decryption and signature generation, i.e., any private-key operations. Also, it is possible for eavesdropper to measure the time during performance of cryptographic operation. And, it is able to analyze how a cryptographic device behaves when certain errors are encountered (Fig. 2).

3.1 Side-Channel Attack in Cryptography

Side-channel attacks of different kinds have existed for many years, so has also general power analysis attacks. More recently, from the late 1990s, another more specific power analysis attack form has evolved. This is the differential power analysis (DPA) attack. This is an attack form which is a non-invasive attack. This means that the attack does not influence the target victim, making it hard or even nearly impossible to detect. There are different forms of DPA attacks, which have different characteristics and qualities. Power analysis attacks are used to extract information about cryptographic keys. There are many devices which are vulnerable against power analysis attacks. The attacks are cheap to perform, requires little computing power, and are relatively easy to understand and perform.

3.2 Types of Side-Channel Attack

A side-channel analysis attack takes advantage of implementation-specific characteristics. It broadly categorized into two categories: active side-channel attack also known as tamper attacks. Another one is passive side-channel attack. Again, the passive attacks are further categorized into two types of attacks. They are simple attacks and differential analysis attacks [10].

3.3 Simple Attacks

In these attacks, attacker can directly guess the secret key using side-channel information. A simple analysis can help attacker to exploit the relationship between executed operations and the side-channel information [10].

3.4 Differential Attacks

This attack exploits relationship between side-channel information and processed data. Here, one hypothetical model is used to guess the rules of side-channel information of device.

3.5 Power Analysis Attacks

This attack is related to power consumption analysis of the unit during encryption operation performance. This attack is again further categorized into simple power analysis attack and differential power analysis attack.

Simple Power Analysis (SPA): Simple power analysis is a technique that involves interpretation of direct power consumption which is collected during encryption and decryption operation. It is based on looking at the visual representation. Paul Kocher and his colleagues, Jaffe and Jun, have done some leading work in the field of power analysis. They have written two papers together on differential power analysis (DPA), thus they describe SPA in both papers. In their first paper [11] published in 1999, SPA is described as such.

SPA can yield information about a devices operation as well as key material. SPA could be used to collect information about the targets cryptographic implementations by, e.g., interpret how many rounds are used during encryption/decryption. SPA is the simplest form of power analysis. Kocher et al. from 2011 state that simple power analysis is a collection of methods for inspection power traces to gain insight into a devices operation, including identifying data-dependent power variations. SPA focuses on examining features that are directly visible in a single power trace evident by comparing pairs of power traces. SPA can, e.g., recover key information by monitoring program flow.

Differential Power Analysis (DPA): DPA is an attack method which is much more powerful attack than SPA. In addition to large-scale power variations found with SPA, DPA searches for correlations between different traces. There are several different DPAs.

Correlation Power Analysis (CPA): The CPA is a form of DPA. It differs a bit from the difference of means attack when searching for correlations. CPA uses a power model. This model is used to say something about the power consumption given a specific plaintext and key combination. CPA attacks have many models for expressing this. The two most common power models are the Hamming weight and the Hamming distance models. attack.

3.6 Fault Attacks

Fault analysis attacks are the recent and more powerful cryptanalysis attack to perform some faulty operations, with the expectation that the results of the fault operation will leak information about the secret key involved.

3.7 Timing Attacks

This attack based on measuring the time it takes for a unit to perform operation. In the timing attacks, one can guessed the key by observing the key combination and how much time it takes to dial from number to number.

4 Counter Measurement of Side-Channel Attack

For power analysis attacks a good signal-to-noise ratio (SNR) is essential when measuring the target MCUs power consumption. With a higher SNR, fewer traces are needed in order to have a successful attack. A trace is a series of samples. For power analysis, a trace must at least contain enough samples to cover the cryptographic operation of interest. SNR can be influenced by counter measures added in both hardware and firmware, as well as bad measuring equipment. When the SNR is good, it is easier to differentiate traces from one another and find needed correlations for the running attacks. In order to achieve good trace measurements, some considerations are needed.

There are various counter measures that exist in the literature for measuring side-channel attack. Some of them are mentioned in the followings [12]:

1. Constant Exponentiation Time: In this counter measure, it must be ensured that all the exponentiation operation consumes same amount time before showing the results. It is a simple fix counter measure. However, it provides degrade performance.
2. Random Delay: It provides better performance than previous one by adding random delay to the exponentiation algorithm. As a result, there is confuse for the timing attack. According to Kocher [12], if defenders do not add enough cryptography and network security noise, the attackers would succeed to collect additional measurements to compensate for the random delays.
3. Blinding: The counter measure blinding is measured by multiplying the ciphertext with a random number before exponentiation operation perform. As a result, the attacker did not guess ciphertext bits which are being processed inside the computer. Thus, it becomes success to prevent the bit-by-bit analysis which is very essential for timing attack.

Besides the above common counter measures, some of the counter measurements against power analysis attacks such as power consumption balancing, reduction of signal size, addition of noise, shielding, and modification of the algorithms design timing attack exist in the literature [13].

5 Research Issues and Future Direction

Based on this literature survey, we have identified the following research issues and challenges.

1. Most of the existing works in the literature are based on AES and DES cryptography algorithm. However, as per the literature concern no work has been reported on improvement of side-channel attack resilience based on ECC algorithm. Whereas, ECC is a very popular algorithm for mobile devices because of its smaller size compared to other cryptographic algorithm.
2. The work reported in the literature had been done in most of the cases based on the same encryption key in every test. In order to improve the results, more (random) encryption keys should be used to find if there are any correlations between key and number of traces needed. Maybe some keys are more resilient than others.
3. The primary focus of Internet of things (IoT) is to remove the gap between physical and virtual world as processing of information is increased day by day through the network. So, the improvement of security of IoT devices also necessary. In this domain, side-channel attacks play a major role to the system in practice such as micro-architectures of processors and their power consumption, and electromagnetic emanation reveals sensitive information to adversaries.

6 Conclusion

Side-channel attacks are major research area in Internet of things (IoT) domain. In this survey paper, different types of Internet of things attacks with special emphasis on side-channel attacks are discussed. This paper also provides the knowledge about IoT attacks taxonomy, existing attacks of side-channel attack, counter major against different side-channel attacks. The research issues in this domain and future direction of research are also highlighted in this paper.

References

1. Ashton, K.: That internet of things thing. RFID J. **22**(7), 97–114 (2009)
2. Akyildiz, I.F., Su, W., Sankarasubramaniam, Y., Cayirci, E.: A survey on sensor networks. IEEE Commun. Mag. **40**(8), 102–14 (2002)
3. Awerbuch, B., Scheideler, C.: Group spreading: a protocol for provably secure distributed name service. In: International Colloquium on Automata, Languages, and Programming 2004, pp. 183–195. Springer, Berlin, Heidelberg
4. Borowik, G., Chaczko, Z., Jacak, W., łuba, T. (eds) Computational Intelligence and Efficiency in Engineering Systems. Springer, Berlin (2015)
5. Kelsey, J., Schneier, B., Wagner, D., Hall, C.: Side channel cryptanalysis of product ciphers. In: European Symposium on Research in Computer Security 1998, pp. 97–110. Springer, Berlin, Heidelberg
6. Deogirikar, J., Vidhate, A.: Security attacks in IoT: a survey. In: 2017 International Conference on I-SMAC (IoT in Social, Mobile, Analytics and Cloud) (I-SMAC), pp. 32–37. IEEE, New York (2017)
7. Forouzan, B.A.: Cryptography & Network Security. McGraw-Hill, New York (2007)
8. Brier, E., Clavier, C., Olivier, F.: Correlation power analysis with a leakage model. In: International Workshop on Cryptographic Hardware and Embedded Systems 2004, pp. 16–29. Springer, Berlin, Heidelberg
9. Coppersmith, D.: The Data Encryption Standard (DES) and its strength against attacks. IBM J. Res. Dev. **38**(3), 243–50 (1994)
10. Standaert, F.X.: Introduction to side-channel attacks. In: Secure Integrated Circuits and Systems 2010, pp. 27–42. Springer, Boston, MA
11. Messerges, T.S.: Using second-order power analysis to attack DPA resistant software. In: International Workshop on Cryptographic Hardware and Embedded Systems 2000, pp. 238–251. Springer, Berlin, Heidelberg
12. Tunstall, M., Mukhopadhyay, D., Ali, S.: Differential fault analysis of the advanced encryption standard using a single fault. In: IFIP International Workshop on Information Security Theory and Practices 2011, pp. 224–233. Springer, Berlin, Heidelberg
13. Prestegrd, H.: Improving side channel attack resilience for IoT devices (Master's thesis, NTNU) (2018)

Optimization of Geotechnical Parameters Used in Slope Stability Analysis by Metaheuristic Algorithms

Geetanjali Lohar, Sushmita Sharma, Apu Kumar Saha, and Sima Ghosh

Abstract Quick development of computer execution empowers new improvements in the field of geotechnical engineering and related zones. Considering design variables, constraints, and objectives in any complex problem, conventional optimization techniques are usually inadequate to find the best solution. So, to overcome these problems, many metaheuristic optimization methods are applied successfully. In this paper, study has been performed to use different methods (DE, BOA, SCA, and PSO) to find the solution of civil engineering problem, i.e., slope stability analysis. Slope stability can be defined as resistance of an inclined soil to withstand or undergo the movement. It is crucial because failure of slope may lead to loss of life, economy, and property. Stability of slopes involves such complexity in case of their different geotechnical parameters. In this paper, different metaheuristic algorithms have been used for optimization of factor of safety and other related parameters of slope stability analysis.

Keywords Slope stability · Factor of safety · Optimization · Metaheuristic algorithm

---

G. Lohar · S. Ghosh
Department of Civil Engineering, National Institute of Technology Agartala, Agartala, Tripura, India
e-mail: lohargeetanjali@gmail.com

S. Ghosh
e-mail: sima.civil@nita.ac.in

S. Sharma · A. K. Saha (✉)
Department of Mathematics, National Institute of Technology Agartala, Agartala, Tripura, India
e-mail: apusaha_nita@yahoo.co.in

S. Sharma
e-mail: snsush01@gmail.com

J. K. Mandal et al. (eds.), *Applications of Internet of Things*, Lecture Notes in Networks and Systems 137, https://doi.org/10.1007/978-981-15-6198-6_21

1 Introduction

Instability of slopes is one of the important concerns in geotechnical engineering. It has been recognized as one of the most frequent natural disaster in mountainous regions that can cause serious injuries, loss of life, economy and property damage. More importantly, it is an incessant cause of suffering because it puts human life in danger. Factors causing instability of slopes are slope geometry, groundwater condition, development of weak zones, properties of slope forming material, structural discontinuity, disruption in geological formation, and heavy rainfall. So, there is a necessity of maximum slope stability and minimum factor of safety which can be obtained by metaheuristic optimization. Slope stability can be defined as resistance of an inclined soil to withstand or undergo the movement.

Many researchers like Terzaghi [1], Newmark [2], Seed [3], Sarma [4], Kramer and Smith [5], Ling et al. [6], Rathje and Bray [7], Choudhury et al. [8], Choudhury and Modi [9] used pseudo-static model, Eskandarinejad and Shafiee [10] Chanda [11] used pseudo-dynamic model and Pain et al. [12] and Chanda et al. [13] proposed modified pseudo-dynamic model for slope stability analysis. The results are computed in using conventional optimization methods. But, nowadays researchers like Nama et al. [14] have used different metaheuristic optimization methods for optimization of geotechnical parameters of retaining wall in which pseudo-static model was used, and Saha et al. [15] have used hybrid symbiosis organisms search algorithm for pseudo-dynamic bearing capacity analysis of shallow strip footing.

In this paper, particle swarm optimization (PSO) [16], differential evolution (DE) [17], butterfly optimization algorithm (BOA) [18] and sine cosine algorithm (SCA) [19]. Algorithm have been used for optimization of geotechnical parameters using modified pseudo-dynamic slope stability analysis. Stability of slopes involves such complexity in case of their different geotechnical parameters due to which use of metaheuristics is preferable.

2 Formulation of the Slope Stability Problem

A slope stability analysis problem as suggested in Chanda et al. [13], in which modified pseudo-dynamic method is used considering a limit equilibrium approach, is selected for the study. The main objective of the study is to maximize the stability of the slope under seismic condition. Forces acting on the slope are shown in Fig. 1, and these forces are used for modified pseudo-dynamic analysis. Different geometrical parameters and components are given in Chanda et al. [13]. In this method of slope stability analysis, the soil medium is considered as Kelvin–Voigt model. Kelvin–Voigt model is one in which solid consists of purely elastic component spring and purely viscous component dashpot which are connected in equivalent to each other that also resist the shear deformation, and this is given by following equation:

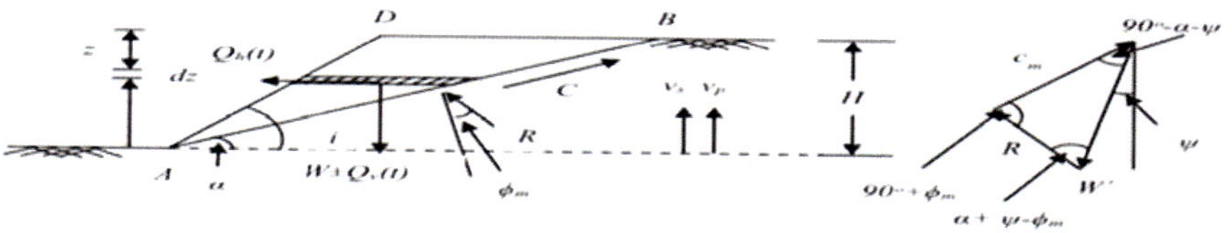

Fig. 1 Failure mechanism of soil wedge and forces acting on a slope

$$\tau = \gamma_s G + \eta \frac{\partial \gamma_s}{\partial t} \tag{1}$$

where τ represents the shear stress, γ_s denotes shear strain, G is the shear modulus, viscosity of the soil is given by η, and t is the time. For damped condition subjected to harmonic loading, viscosity is given by $\eta = 2G\xi\omega$ where $\xi =$ damping ratio and ω is the angular frequency of the shear wave motion.

In this model, propagation of wave is considered along the z-axis and solution of equation of motion of Kelvin–Voigt model is given as follows:

$$\rho \frac{\partial^2 u_h}{\partial t^2} = G \frac{\partial^2 u_h}{\partial z^2} + \eta \frac{\partial^3 u_h}{\partial z^2 \partial t} \tag{2}$$

$$\rho \frac{\partial^2 u_v}{\partial t^2} = (\lambda + 2G) \frac{\partial^2 u_v}{\partial z^2} + (\eta_1 + \eta_s) \frac{\partial^3 u_v}{\partial z^2 \partial t} \tag{3}$$

where ρ is density of the soil, λ is lame constant, u_h and u_v horizontal and vertical displacements.

For further solution of these equations, by applying boundary conditions such as height of slope and shear stress at free surface $z = 0$, both horizontal and vertical displacement is evaluated. Displacement and acceleration equation given by Chanda et al. [13] are as follows:

$$u_h(z,t) = \frac{u_{h0}}{m_s^2 + n_s^2}\left[(m_s m_{sz} + n_s n_{sz})\cos(\omega_s t) + (n_s m_{sz} - m_s n_{sz})\sin(\omega_s t)\right] \tag{4}$$

In which,
base displacement:

$$u_b = u_{h0} e^{i\omega t} \tag{5}$$

$$l_{s1} = \frac{\omega_s H}{V_s}\left\{\frac{\sqrt{(1+4\zeta^2)} + 1}{2(1+4\zeta^2)}\right\}^{0.5} \tag{6}$$

$$l_{s2} = \frac{\omega_s H}{V_s}\left\{\frac{\sqrt{(1+4\zeta^2)} - 1}{2(1+4\zeta^2)}\right\}^{0.5} \tag{7}$$

$$m_s = \cos(l_{s1})\cosh(l_{s2}) \tag{8}$$

$$n_s = -\sin(l_{s1})\sinh(l_{s2}) \tag{9}$$

$$m_{sz} = \cos\left(\frac{l_{s1}z}{H}\right)\cosh\left(\frac{l_{s2}z}{H}\right) \tag{10}$$

$$n_{sz} = -\sin\left(\frac{l_{s1}z}{H}\right)\cosh\left(\frac{l_{s2}z}{H}\right) \tag{11}$$

For acceleration, we have to differentiate the above Eq. (4) twice with respect to time and substituting $k_h g = -\omega_s^2 u_{h0}$. Horizontal acceleration according to expressed as;

$$a_h(z,t) = \frac{k_h g}{m_s^2 + n_s^2}\left[(m_s m_{sz} + n_s n_{sz})\cos(\omega_s t) + (n_s m_{sz} - m_s n_{sz})\sin(\omega_s t)\right] \tag{12}$$

Similarly, for vertical displacement and acceleration, primary wave is considered for formulation of equations. So, horizontal displacement is given as:

$$u_v(z,t) = \frac{u_{v0}}{m_p^2 + n_p^2}\left[\left(m_p m_{pz} + n_p n_{pz}\right)\cos\left(\omega_p t\right) + \left(n_p m_{pz} - m_p n_{pz}\right)\sin\left(\omega_p t\right)\right] \tag{13}$$

$$a_V(z,t) = \frac{k_v g}{m_p^2 + n_p^2}\left[\left(m_p m_{pz} + n_p n_{pz}\right)\cos\left(\omega_p t\right) + \left(n_p m_{pz} - m_p n_{pz}\right)\sin\left(\omega_p t\right)\right] \tag{14}$$

2.1 Computation of Stability Number for Design of Slope

A slope of height *'H'* inclined at an angle *'i'* with respect to horizontal is placed with *'c*—ϕ' soil. Failure wedge AB is assumed to be inclined at an angle 'α' with the horizontal. Figure 1 shows forces acting on slope. '*W*' is the self-weight of the soil acting vertically downward, cohesive force '*C*' acts along the failure surface, '*R*' is reaction which acts perpendicular to the failure plane inclined at an angle '$\phi_{m,}$' and 'Q_h' and 'Q_v' are horizontal and vertical seismic inertia which acts in horizontal and vertical direction.

Consider a thin element of wedge at depth (*z*), thickness (d*z*), mass—*m*(*z*). Horizontal and vertical seismic inertia force is computed as;

$$Q_h(\mathrm{z}, \mathrm{t}) = 0.5k_h\gamma(\cot\alpha - \cot i)H^2\left[\frac{m_s s_1 + n_s s_1}{m_s^2 + n_s^2}\cos(\omega t) + \frac{n_s s_1 + m_s s_2}{m_s^2 + n_s^2}\sin(\omega t)\right] \tag{15}$$

$$Q_V(\mathrm{z}, \mathrm{t}) = 0.5k_v\gamma(\cot\alpha - \cot i)H^2\left[\frac{m_p p_1 + n_p p_1}{m_p^2 + n_p^2}\cos(\omega t) + \frac{n_p p_1 + m_p p_2}{m_p^2 + n_p^2}\sin(\omega t)\right] \tag{16}$$

The absolute weight of wedge due to its self-weight and effect of seismic inertia forces is given by,

$$W' = \sqrt{Q_h^2 + (W \pm Q_v)^2} \tag{17}$$

Considering the force triangle as shown in Fig. 1 and applying sine rule, stability number for modified pseudo-dynamic condition is calculated as,

$$S_n = \frac{\sin\alpha\sin(\alpha + \psi + \phi_m)(\cot\alpha - \cot i)}{2\cos\phi_m}\left\{(1 + k_h^2)a^2 \pm 2k_v b + k_v^2 b^2\right\}^{0.5} \tag{18}$$

where

$$a = \left[\frac{m_s s_1 + n_s s_1}{m_s^2 + n_s^2}\cos(\omega t) + \frac{n_s s_1 + m_s s_2}{m_s^2 + n_s^2}\sin(\omega t)\right] \tag{19}$$

$$b = \left[\frac{m_p p_1 + n_p p_1}{m_p^2 + n_p^2}\cos(\omega t) + \frac{n_p p_1 + m_p p_2}{m_p^2 + n_p^2}\sin(\omega t)\right] \tag{20}$$

3 Optimization Algorithms

3.1 Particle Swarm Optimization

PSO is a swarm-intelligent-based technique which is used for optimization of continuous nonlinear functions. It is inspired from the social and individual behavior of schools of fishes or flocks of birds in foraging. In this method, firstly set of random solution is created, and on the basis of the best solution that is obtained or swarm has found, velocity of a particle is updated. At every iteration, best solutions are saved in the PSO algorithm, and because of this reason, there is always huge chance of finding enhanced solutions.

3.2 Differential Evolution

Differential evolution (DE) is a population-based stochastic search algorithm which consists of three operators: mutation, crossover, and selection. DE utilizes number of population parameter vectors (NP), and it is a parallel direct search method.

Mutation—This phase involves creation of new parameter vector by including a weighted contrast vector between two populace individuals to a third part.

Crossover—The term crossover refers to collaborating of parameters. If the trial vector values are smaller than the objective vector, the objective vector is replaced by trial vector in the next generation.

Selection—During selection stage, greedy criteria is used, and then, it decides whether vector should become a member of generation.

3.3 Butterfly Optimization Algorithm

Butterfly optimization algorithm is a metaheuristic algorithm inspired from nature which is mainly based on food garging behavior of butterflies. In BOA, butterfly is considered to be the search agents and assumed that they generate the fragrance with some amount of intensity. The fragrance is then compared with the fitness of butterfly. When butterfly will migrate from one place to another, fitness of butterfly will vary accordingly. The fragrance which is generated propagates over a particular distance which is sensed by all the butterflies present nearby. Butterflies sense this propagated fragrance, and a social knowledge network is developed. Butterfly moves a step forward toward best butterfly by smelling the fragrance of it, and this phenomenon is termed as global search phase of BOA. Similarly, if butterfly is not able to smell the fragrance, it will take random steps, and then, it is termed as local search phase of BOA.

This idea of detecting depends on the three parameters, i.e., intensity of stimuli (I), sensor modality, (c) and power exponent (a). The method by which the utilization of vitality is estimated and prepared by the sensors is called as sensor modality. The magnitude of the stimulus is given by I, and this magnitude is correlated with the fitness of butterfly. The phenomenon of searching for food and mating partner by butterflies can be occurred by both local and global scale.

3.4 Sine Cosine Algorithm

Sine cosine algorithm is a populace-based algorithm which is used for determining optimization problems. The purpose of sine cosine algorithm (SCA) is to permit the candidate solutions to fluctuate toward and away from the finest solution by creating various number of initial random candidate solution. This is satisfied by using the mathematical approach which is governed by trigonometric sine and cosine functions. Random set is evaluated frequently by an objective function and upgraded by set of tenets that is center of an advancement strategy. Population-based algorithm search for optima stochastically, because of which there is no guarantee of finding an answer in single run. Increase in adequate number of iterations and random solutions, probability of receiving a global optimum increases. In the exploitation stage, there are steady changes in the arbitrary arrangement (random solutions) and random variables, which are extensively not exactly those in the investigation stage.

4 Results and Discussion

The objective function is stability number S_n, which should be maximized. It is an unconstrained problem. All the terms are constant except α and t/T. So, S_n is optimized for different values of α and t/T ranging from 0° to 90°. Optimization is obtained by using the following methods: DE, PSO, SCA, and BOA. Optimal values of S_n are shown in Table 1, and results obtained from different methods are compared (Table 2).

MATLAB, R2017a is used for computation of stability number by varying α and t/T optimum value. Results for the seismic stability of slope are presented in tabular forms. Four optimization algorithms, namely DE, PSO, BOA, and SCA, are used to analyze stability number, and the results are presented in tabular form. Among all the four algorithms, BOA shows better results in most of the cases which means it gives maximum stability and minimum factor of safety (Fig. 2).

Table 1 Stability number obtained from different optimization methods and compared with Chanda et al. [13] using $\zeta = 20\%$, $k_h = 0.3$, $k_v = k_h/2$, $i = 20–90°$, $\Phi = 30°$

i	DE	BOA	PSO	SCA	Chanda et al.
20	0.0142	0.0168	0.0421	0.0152	0.014
30	0.0432	0.0485	0.0852	0.0467	0.041
40	0.0736	0.0856	0.0739	0.0763	0.072
50	0.1125	0.1267	0.1095	0.1154	0.107
60	0.1564	0.1721	0.1326	0.1679	0.145
70	0.1930	0.2230	0.1750	0.2045	0.18
80	0.2404	0.2804	0.2563	0.2562	0.237
90	0.3082	0.3482	0.3124	0.3291	0.204

Table 2 Stability number obtained from different optimization methods and compared with Chanda et al. [13] using $\zeta = 20\%$, $k_h = 0.3$, $k_v = k_h/2$, $i = 30–40°$, $\Phi = 20°, 30°, 40°$

i	Φ	DE	BOA	PSO	SCA	Chanda et al.
30	20	0.0914	0.1014	0.1042	0.0693	0.086
	30	0.0432	0.0485	0.0449	0.0467	0.041
	40	0.0110	0.0130	0.0245	0.0097	0.011
40	20	0.1191	0.1391	0.1162	0.1076	0.117
	30	0.0736	0.0856	0.0739	0.0763	0.072
	40	0.0345	0.0427	0.3670	0.0345	0.036

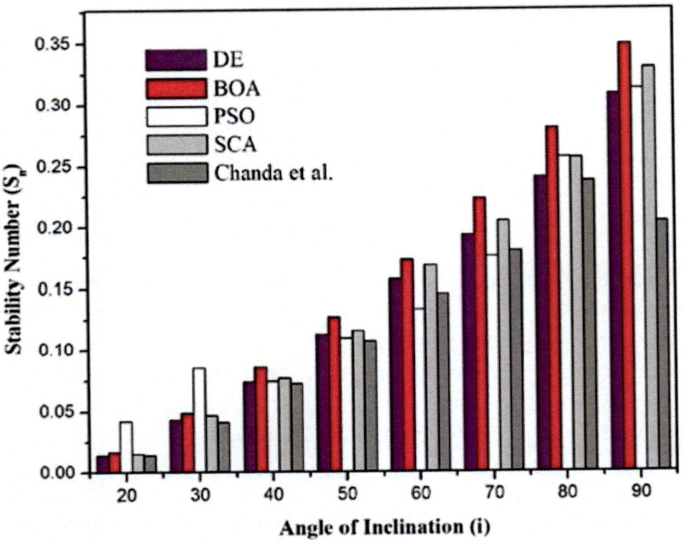

Fig. 2 Comparison of results obtained for $\zeta = 20\%$, $k_h = 0.3$, $k_v = k_h/2$, $i = 30–40°$, $\Phi = 20°$, $30°$, $40°$

5 Conclusions

Optimization problems can be effectively used to solve civil engineering problems. Although it has been used for solving many geotechnical problems, the concept is new for modified pseudo-dynamic analysis. So, an attempt has been made to carry out slope stability analysis in which modified pseudo-dynamic method is used, and four optimization algorithms, namely DE, PSO, BOA, and SCA, are used to analyze stability number, and the results obtained are compared with Chanda et al. All the methods give better results as compared to Chanda et al. Among all the four methods, BOA gives best results which means, BOA is more effective to solve slope stability problem for modified pseudo-dynamic conditions.

References

1. Terzaghi, K.: Mechanisms of Landslides. Geological Society of America, Engineering Geology, Berkeley (1950)
2. Newmark, N.: Effects of earthquakes on dams and embankments. Geotechnique **15**(2), 139–160 (1965)
3. Seed, H.B.: A method for the earthquake resistant design of earth dams. J. Soil Mech. Found. Div. ASCE **92**(SM1), 13–41 (1966)
4. Sarma, S.K.: Seismic stability of earth dams and embankments. Geotechnique **25**, 743–761 (1975)
5. Kramer, S.L., Smith, M.W.: Modified Newmark model for Seismic displacements of compliant slopes. J. Geotech. Geoenviron. Eng. **123**(7), 635–644 (1997)
6. Ling, H.I., Mohri, Y., Kawabata, T.: Seismic analysis of sliding wedge: extended Francais–Culmann's analysis. Soil Dyn. Earthquake Eng. **18**(5), 387–393 (1999)
7. Rathje, E.M., Bray, J.D.: Nonlinear coupled seismic sliding analysis of earth structures. J. Geotech. Geoenviron. Eng. **126**(11), 1002–1014 (2000)
8. Choudhury, D., Basu, S., Bray, J.D.: Behaviour of slopes under static and seismic conditions by limit equilibrium method, Denver, Colorado: Proceedings of Geo-Denver (2007)
9. Choudhury, D., Modi, D.: Displacement based seismic stability analysis of reinforced and unreinforced slopes using planner failure surfaces. In: Geotechnical Earthquake and Engineering and Soil Dynamics IV Congress, ASCE, pp. 1–10 (2008)
10. Eskandarinejad, A., Shafiee, A.H.: Pseudo-dynamic analysis of seismic stability of reinforced slopes considering non-associated flow rule. J. Central South Univ. Technol. **18**, 2091 (2011)
11. Chanda, N.: Pseudo-dynamic analysis of slope. Int. J. Adv. Res. Sci. Eng. **4**(1), 729–736 (2015). ISSN-2319-8354(E)
12. Pain, A., Choudhury, D., Bhattacharya, S.K.: Seismic stability of retaining wall-soil sliding interaction using modified pseudo-dynamic method. Geotech. Lett. **5**, 56–61 (2015)
13. Chanda, N., Ghosh, S., Pal, M.: Analysis of slope using modified pseudo-dynamic method. Int. J. Geotech. Eng. (2017)
14. Nama, S., Saha, A.K., Ghosh, S.: Parameters optimization of geotechnical problem using different optimization algorithm. Geotech. Geol. Eng.: Int. J. (2015)
15. Saha, A., Saha, A.K., Ghosh, S.: Pseudodynamic bearing capacity analysis of shallow strip footing using the advanced optimization technique hybrid symbiosis organisms search algorithm with numerical validation. In: Advances in Civil Engineering (2018)
16. Kennedy, J., Eberhart, R.: Particle Swarm Optimization. IEEE0-7803-2768-3/95/4.000 (1995)
17. Storn, R., Price, K.: Differential evolution—a simple and efficient adaptive scheme for global optimization over continuous spaces. J. Glob. Optim. (2019)
18. Arora, S., Singh, S.: Butterfly optimization algorithm: a novel approach for global optimization. Soft Comput **23**, 715–734 (2019). https://doi.org/10.1007/s00500-018-3102-4
19. Mirjalili, S.: SCA: A sine cosine algorithm for solving optimization problems. Knowledge-Based Systems (2016)

An Improved ANN Model for Prediction of Solar Radiation Using Machine Learning Approach

Rita Banik, Priyanath Das, Srimanta Ray, and Ankur Biswas

Abstract An accurate forecast of weather is essential for obtaining energy from Renewable sources. The objective of this paper is to present an analysis of weather parameters and comparison among different models of the weather prediction from accessible parameters and finally deriving a new technique for solar prediction in support of photovoltaic output power. Artificial Neural Network model with 5 weather parameters from NASA POWER dataset have been utilized to predict the day-ahead solar radiation and evaluated against real data measured for 4 years at Agartala, India (Latitude 23.83° N and Longitude 91.282° E). Results detailed in this work confirm the best predicting potential of the proposed method. The proposed model has been shown to predict solar radiation with accuracy of 83% shows the robustness of the system.

Keywords Weather forecast · ANN · Photovoltaic · Solar radiation · Machine learning

1 Introduction

Photovoltaic (PV) has turned to be a popular and a green energy among the available renewable energy sources. Governments in various countries have chosen feed-in

R. Banik (✉) · P. Das · S. Ray
National Institute of Technology Agartala, Agartala, Tripura, India
e-mail: rita.nit@gmail.com

P. Das
e-mail: priyanathdas@yahoo.co.in

S. Ray
e-mail: srimanta.chemical@nit.ac.in

A. Biswas
Tripura Institute of Technology, Narsingarh 799009, Tripura, India
e-mail: abiswas.tit@gmail.com

J. K. Mandal et al. (eds.), *Applications of Internet of Things*, Lecture Notes in Networks and Systems 137, https://doi.org/10.1007/978-981-15-6198-6_22

tariffs to tackle the setback of the greenhouse gas emissions [1]. Hence, the installation of PV panels in the houses of people has become simple. In this context, prediction of PV power provoked as a research domain. The crisis of PV power prediction for smart-grid and microgrid operations has been analyzed with a minute prediction time horizon [2]. PV power prediction is utilized for designing few strategies in bidding market, and sizing and defining storage systems that can be built with PV plants. However, the accurateness of the prediction has developed to be exceptionally essential [3]. There are various approaches in regard to prediction: the physical or parametric method, statistical or black box method, and hybrid model or gray box. Recently, computational intelligence techniques are preferred in resolving problems of optimization, forecasting, sizing, and control of stand-alone, grid-connected, and hybrid photovoltaic systems [4–6]. Machine learning techniques like artificial neural networks (ANNs) appear to be another promising PV power prediction tool in particular [7–11]. The prediction of PV generation is conventionally deterministic. But this technique is unable to produce information related to prediction error margins and the self-assurance of the prediction [12]. A probabilistic approach indicates the most possible values of the power generation and estimates wide ranges of probable value and the likelihood linked to it. It is also helpful in every actions requiring risk and ambiguity management like network load balance and import/export of energy.

The objective of the paper, in line with earlier algorithms, is to improvise the prediction of solar radiation for PV power output performed one day in advance through five parameters using ANN. Typically, accuracies in the weather prediction robustly involve the PV prediction techniques. The novelty of this effort is that the proposed techniques are evaluated with the similar observed data to check the accuracies in line with the theory of the best option of weather forecasting models. Later, the unchanged forecasting methodology is considered with same weather prediction system under standard conditions. This paper is pre-arranged as follows: in Sect. 2, the detail about the forecasting framework is detailed; in Sect. 3, the methodologies are presented; in Sect. 4, the experimental result is shown; and lastly, in Sect. 5, the concluding remarks are commented.

2 Prediction Framework

2.1 Dataset

The dataset is obtained from National Aeronautics and Space Administration (NASA) POWER Project Merra2 [13] (https://power.larc.nasa.gov/) for latitude (decimal degrees): 23.83 N and longitude (decimal degrees): 91.282 E in hourly resolution for nine attributes from January 1, 2016, to September 23, 2019.

2.2 *Feature Extraction*

Multiple factors influence the future PV power output like irradiance, temperature, pressure, humidity, wind speed, cloud coverage, etc. Some active methodology like ARIMA and GM (1,1) depend on the pattern of historical time series to calculate PV power devoid of taking into account the persuade of different external characteristics. These schemes encompass the apparent shortcoming that it is hard to adjust to the environment changeability, particularly for the modulation point. Hence, to improve the accuracy of forecasting the PV power output, the effect of different features considering is essential. But, the non-related parameters or characteristics should not be placed in the model because it increases the model complexity which interferes with other parameters of the model. Therefore, feature extraction is obligatory prior to building the actual model.

2.3 *Cluster Analysis*

When feature values are dissimilar, the output result is also dissimilar. Data clustering as per the feature resemblance is obliging in order to improve the accuracy. Here, the training set is split into numerous groups and further analyzed for improving the accuracy. Various established clustering algorithms are available whose outcome relies on the application area and dataset. The popular K-means technique is among classical clustering algorithms that has fast convergence and good stability [14] for prediction with superior stoutness. It has been utilized during data pre-processing stage. In this paper, the training set has been grouped by means of K-means algorithm with an intention to keep the characteristics of the each group identical.

3 Methodology

This section describes the methodology adopted for designing the proposed model of prediction. The overall framework is shown in Fig. 1.

3.1 *Artificial Neural Network*

ANN is extensively used to address problems related to weather prediction. It is a data-driven model that resembles the structure of a human brain neural network which is based on the perceptron (to combine the task of neuron and recognition). It contains one layer each for input and output layer containing nodes for data operations. It can be extended to multilayer structure with addition of a hidden layer and nodes within

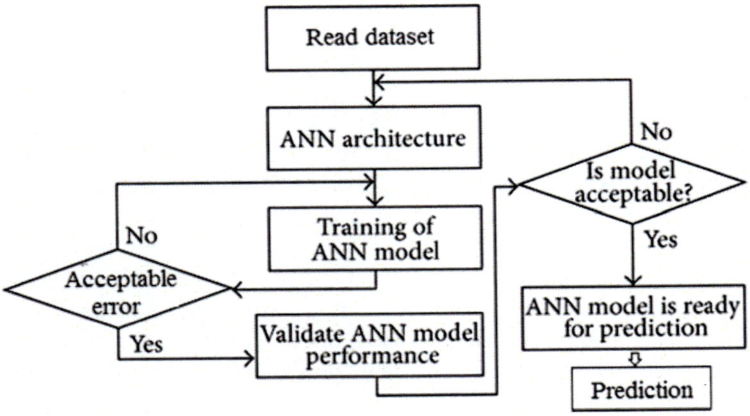

Fig. 1 Framework of the proposed approach

input as well as output layer. A three-layered feed-forward neural network is popular for building forecasting models [15–18]. The data from the input layer is passed on to each neuron in the hidden layer which is further transferred to output layer via series of operations. A neural network with three layers is represented by a linear combination of the transferred input values as,

$$y' = f_0\left[\sum_{j=1}^{n} w_{kj} \cdot f_h \cdot \left(\sum_{j=1}^{m} w_{ji} \cdot x_i + w_{jb}\right) + w_{kb}\right] \quad (1)$$

Here y' represents the output predicted, f_0 being output neuron activation function with n representing the output number. The weight linking the jth neuron of hidden layer and kth neuron of output layer is denoted by $w_{kj,}$ and the weight involving the ith neuron of input layer and jth neuron in the hidden layer is denoted by w_{ji}. f_h and m represents activation function and number of hidden neurons, respectively. x_i represents the ith input variables while w_{jb} and w_{kb} symbolize the bias for the jth hidden neuron and kth output neuron, respectively (Fig. 2).

ANN model learning involves training method that requires searching the optimal weight of Eq. 1. The weight minimizing E (sum of error) of the neural network in Eq. 2 was computed through back-propagation,

$$E = \sum_{n=1}^{N} (y - y')^2. \quad (2)$$

where y is the actual output and y' predicted output.

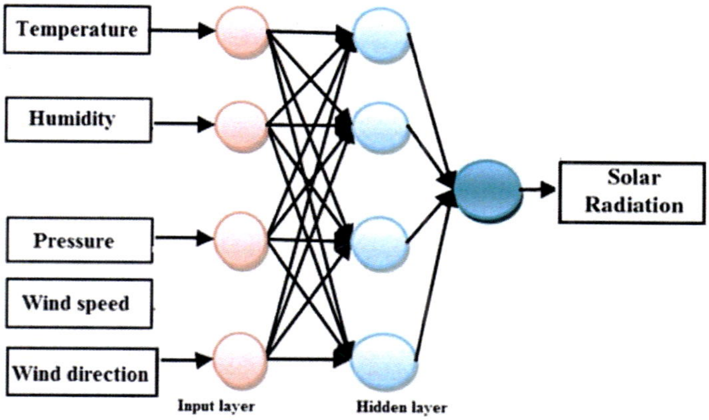

Fig. 2 ANN model

3.2 ANN Model Development

The number of input variables is equivalent to the number of input nodes. One hidden layers has been selected with 0.01 learning rate with hyperbolic tangent sigmoid activation function in hidden layer. The fivefold cross-validation (CV) procedure was utilized to assess the performance of mode and avoid overfitting.

3.3 Epochs

Epoch in neural network expressions is one forward pass and one backward pass of all the training set. Total epochs describe the number of times the algorithm will work through the complete dataset of training. In every epoch, the training samples have a prospect to revise the parameters of model.

4 Results

This section shows the result of the proposed model that will predict the solar radiation from correlation between weather variables using ANN. The parameters used for model are loss 'mse,' optimizer 'adam,' epoch '5,' batch_size '32,' etc. The highest accuracy given by our proposed system is all most 83.32% for five epochs during training phase, and during testing phase of the machine, it has shown accuracy rate near to 74% for five epochs. The number of epoch here means how many times the system is trained or tested with a train dataset and validation dataset, respectively.

Fig. 3 Pearson correlation heatmap

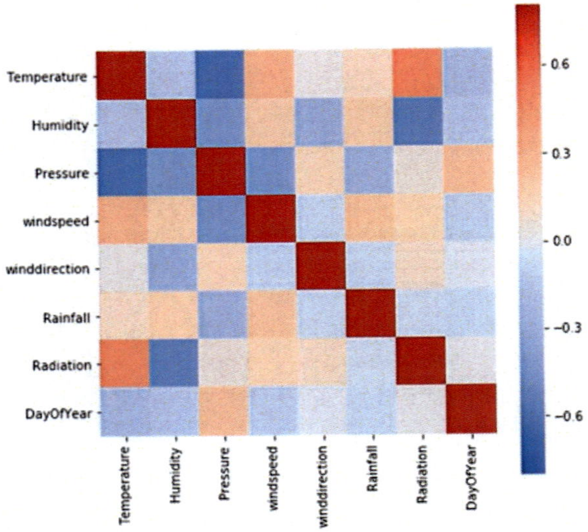

During data analysis, it is clear that temperature has strong correlation with solar radiation. Relationships between pressure/humidity and solar radiation are less clear, but it does appear that humidity has a negative correlation with solar radiation, temperature, and pressure. As expected, solar radiation and temperature both peak at approximately 12:00. Additionally, monthly means of both solar irradiance and temperature appear to decrease as winter approaches, with the exception of a very slight increase in solar radiation from September to October. To visualize the relationships between the variables, a Pearson correlation heat map was plotted as shown in Fig. 3. It is observed that solar irradiance does not have a linear correlation with 'DayOfYear.' Hence, it is excluded from training and prediction.

4.1 Separating the Independent and Dependent Variables

All recorded meteorological variables, except solar irradiance, were included in the independent variables. 'DayOfYear' and 'TimeOfDay(s)' were selected to represent date and time. This would ensure that no problems were encountered if predictions for another year were to be made. Solar irradiance was of course set as the independent variable. Following are set of train (X) and prediction (y) dataset.

X = dataset ['Temperature', 'Pressure', 'Humidity', 'winddirection', 'windspeed', 'DayOfYear', 'TimeOfDay(s)'].

y = dataset ['Radiation'].

The dataset was subsequently split into a training and test set, with an 80% and 20% split, respectively.

Table 1 Feature parameters with scores

Sl	Features	r^2-Score
0	Temperature, pressure, humidity, wind direction	0.836106
1	Temperature, pressure, humidity, wind speed	0.824142
2	Temperature, pressure, wind speed	0.747872
3	Temperature, pressure, humidity	0.674291
4	Temperature, humidity	0.530672

4.2 Feature Selection

Although linear regression can be used to estimate the importance of different features, but may not suitable for nonlinear data. Hence, this attribute was used to perform a backwards elimination procedure, where the least important feature of the regression model was repeatedly removed and the r^2 scores, from cross-validation, of each model were recorded. The features and r^2 scores are given in Table 1

From the dataframe output, it can be seen that model performance stays relatively constant until 'humidity' and 'wind speed' is removed, leaving 'temperature' and 'pressure' as the only features. Without performing any parameter tuning, it appears that the proposed ANN model fit to 'temperature,' 'pressure,' 'humidity,' and 'wind direction' is able to achieve an r^2 score as high as 0.83.

4.3 Result of Cross-validation

Cross validation, with a greater number of folds, i.e., 10, again shows an r^2 score of 0.83.

4.4 Predicting the Test Set

The trained model is the used to predict and test set data, which was not involved in the training process. Explained variance, mean squared error, and r^2 scores were output to evaluate the accuracy of the models predictions.

```
explained variance = 0.8364823038223047

mse = 2.11

r2 = 0.8364491397501871
```

The model loss and epoch-accuracy graph while training with various number of epochs, i.e., 0–5 is shown in Fig. 4.

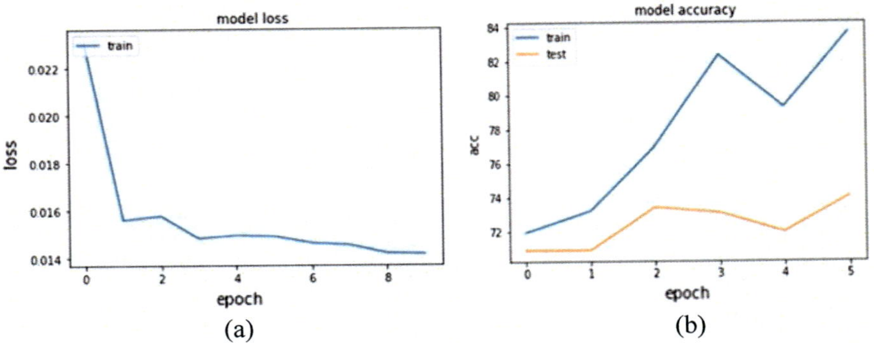

Fig. 4 **a** Model loss and **b** epoch versus accuracy graph (epochs 0–5)

A comparison of actual solar radiation and predicted value for different dates is shown in Fig. 5.

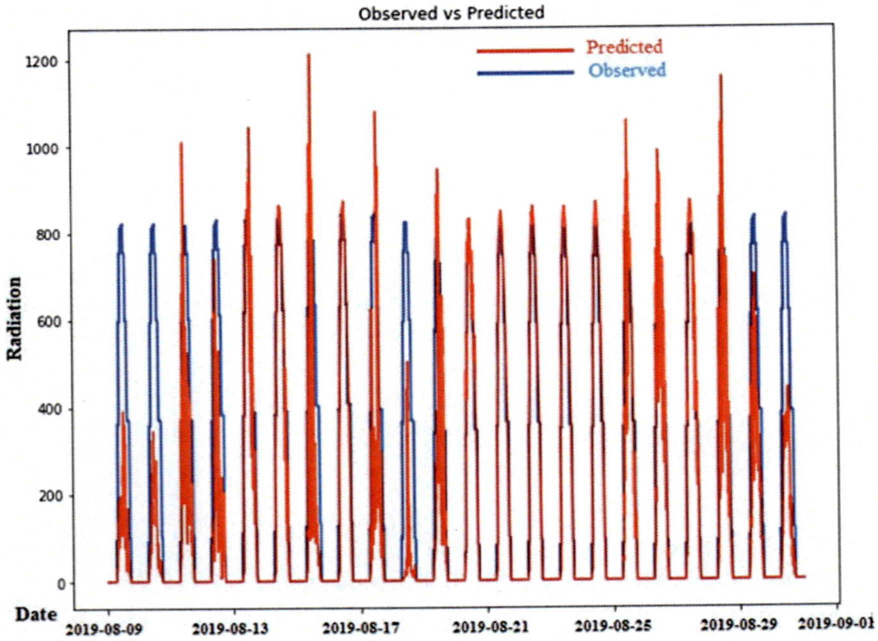

Fig. 5 Comparison of actual and predicted value

Table 2 Comparison of different models

Model	R^2-score	RMSE
Linear regression	0.698	7.3412
Decision tree	0.763	6.4363
Boosted decision tree	0.813	6.2214
ANN (proposed)	0.836	5.4341

4.5 Time Complexity Analysis

The proposed system was implemented on Windows GPU environment with intel CPU (i5, 2.2 GHz) using Python, Keras, and Tensorflow. The average training time for four years of data for each epoch is approximately 5 s. A comparison of the different model in terms of R^2 and RMSE for prediction of solar radiation is given in Table 2.

5 Conclusion and Future Scope

The current active fossil fuels will not be able satisfy the ever-increasing demand for energy; hence, utmost attention has been given to renewable sources of energy. However, the intermittency and varied characteristics of renewable energy like solar and wind have caused a need forecasting model with high accuracies. Therefore, this paper is a contribution to the expansion of an improved solar radiation forecasting technique. In this paper, a novel methodology is shown for prediction of day-ahead solar radiation. The proposed model incorporates the main features of artificial neural network along with easy accessible parameters. The primary objective of this model is to improvise the weather forecasts with the optimized parameters. Initially, time series data from 2016 to 2019 were engaged to analyze the diverse algorithms while training in terms of total days in the dataset. In terms of accuracy, it is revealed that ANN outperforms other algorithms

Future enhancements for recuperating this work are associated with data collection from other locations with diverse weathered conditions. This task is intended to enhance assessment of the prediction competence of the proposed method.

References

1. Price, L., Michaelis, L., Worrell, et al.: Sectoral trends and driving forces of global energy use and greenhouse gas emissions. Mitig. Adapt. Strateg. Global Change **3**, 263–319 (1998)
2. Wan, C., Zhao, J., Song, Y., et al.: Photovoltaic and solar power forecasting for smart grid energy management. CSEE J. Power Energy Syst. **1**(4), 38–46 (2015)

3. Yang, D., Kleissl, J., Gueymard, C.A., et al.: History and trends in solar irradiance and PV power forecasting: a preliminary assessment and review using text mining. Sol. Energy **168**, 60–101 (2018)
4. Antonanzas, J., Osorio, N., Escobar, R., et al.: Review of photovoltaic power forecasting. Sol. Energy **136**, 78–111 (2016)
5. Raza, M.Q., Nadarajah, M., Ekanayake, C.: On recent advances in PV output power fore-cast. Sol. Energy **136**, 125–144 (2016)
6. Inman, R.H., Pedro, H.T.C., Coimbra, C.F.: Solar forecasting methods for renewable energy integration. Prog. Energy Combust. Sci. **39**, 535–576 (2013)
7. Graditi, G., Ferlito, S., Adinolfi, G.: Comparison of photovoltaic plant power production prediction methods using a large measured dataset. Renew. Energy **90**, 513–519 (2016)
8. Dong, Y., Jiang, H.: Global solar radiation forecasting using square root regularization-based ensemble. Mathematical Problems in Engineering, Article ID 9620945 (2019). https://doi.org/10.1155/2019/9620945
9. Hameed, W.I., Sawadi, B.A., Al-Kamil, S.J. et al.: Prediction of solar irradiance based on artificial neural networks. Inventions **4**, 45 (2019). https://doi.org/10.3390/inventions4030045
10. Basaran, K., OzCift, A., Kilinc, D.: A new approach for prediction of solar radiation with using ensemble learning algorithm. Arab. J. Sci. Eng. **44**, 7159–7171 (2019). https://doi.org/10.1007/s13369-019-03841-7
11. Sikiru, S.: Modeling of solar radiation using artificial neural network for renewable energy application. IOSR J. Appl. Phys. **10** (2018). https://doi.org/10.9790/4861-1002030612
12. Bacher, P., Madsen, H., Nielsen, H.A.: Online short-term solar power forecasting. Sol. Energy **83**, 772–783 (2009)
13. Global Modeling and Assimilation Office. MERRA- 2 tavg1_2d_slv_Nx: 2d,1-hourly,time-averaged, single-level, assimilation, single-level diagnostics V5.12.4, Greenbelt, MD, USA, Goddard Earth Sciences Data and Information Services Center (GES DISC) (2015). https://doi.org/10.5067/vjafpli1csiv. Accessed 06 Dec 2019
14. Xu, J., Liu, H.: Web user clustering analysis based on K Means algorithm. In: Proceedings of 2010 International Conference on Information, Networking and Automation (ICINA), Kunming, China (2010), pp. V2-6–V2-9
15. Dawson, C.W., Wilby, R.L.: Hydrological modelling using artificial neural networks. Prog. Phys. Geogr. **25**, 80–108 (2001)
16. De Vos, N.J., Rientjes, T.H.M.: Constraints of artificial neural networks for rainfall-runoff modelling: trade-offs in hydrological state representation and model evaluation. Hydrol. Earth Syst. Sci. **9**, 111–126 (2005)
17. Sumi, S.M., Zaman, M.F., Hirose, H.: A rainfall forecasting method using machine learning models and its application to the Fukuoka city case. Int. J. Appl. Math. Comput. Sci. **22**, 841–854 (2012)
18. Singh, S.K., Jain, S.K., Bardossy, A.: Training of artificial neural networks using information-rich data. Hydrology **1**, 40–62 (2014)

User Behaviour Analysis from Various Activities Recorded in Social Network Log Data

Krishna Das and Smriti Kumar Sinha

Abstract Social network user behaviour analysis is to define behaviour formally in an appropriate manner. This formal behaviour representation helps in finding out the appropriate behaviour pattern from huge social network data sets and to provide a perfect qualitative analysis out of the computed results. People so far have tried to define the composition of user behaviour in terms of set of activities, patterns, way of participation, influence, etc., in the social network. Various methods were employed in characterization of user behaviour in various social network platforms. In this paper, we tried to describe silent user behaviour in social networks. User behaviour can be analysed based on the various silent activities it has performed without directly leaving any footprint in the network. These are regarded as silent behaviour as these are computed from the user generated log data. Types of required data sets, necessary parameter computation and finally analysis of silent behaviour based on the physical significance of the computed parameters are presented here in this paper.

Keywords Silent behaviour · Log data · Network browsing · Activity frequency

1 Introduction

Behaviour term itself is qualitative in literature. So, behaviour description of an user in social network depends on types of activities one has performed by him/her in a specific time frame. This is a very complex representation as real human behaviour greatly differs from social network user behaviour which is again varies from one social network platform to another. Same user in different social network platforms exhibits various activity patterns which again vary from user to user that shapes the

K. Das (✉) · S. K. Sinha
Department of Computer Science and Engineering,
Tezpur University, Napaam, Tezpur, Assam 784028, India
e-mail: krishnadas.mca@gmail.com

S. K. Sinha
e-mail: smritikumarsinha@gmail.com

J. K. Mandal et al. (eds.), *Applications of Internet of Things*, Lecture Notes in Networks and Systems 137, https://doi.org/10.1007/978-981-15-6198-6_23

general user behaviour. Moreover, characterization of behaviour in a computable form to model the general behaviour poses difficulty in terms of processing complexity, exact data representation for the model input and so on.

Again, apart from activity, users may have different types of influences on the network depending on their corresponding structural position in the network. Various methods were employed in characterization of user behaviour in various social network platforms. In this work, mainly silent user behaviour are described. Position of an user in a social network tells many things about his activities, influence, etc. This is called structural behaviour in this study. Another type of behaviour is analysed based on the various activity it has performed. These types of behaviours are called as silent behaviour as it is computed from the user generated log data.

Behaviour in social network platform is regarded as set of patterns of activities performed by a user. Basically, there are two types of activities performed by a user in social network. One is explicit activity such as who has sent messages to whom, who has updated his profile status, uploading photographs. Such types of activities are directly computable from the data contents created by the user. Activity pattern of this category leads to structural behaviour [1–3].

Some activity patterns are very difficult to identify simply by looking at users profile data. These includes browsing others profiles, remain silent spectator in the network, login and logout activity, doing malicious activity such as hijacking someones administrative privilege. These types of activities can be captured only from user log data. Mainly three types of behaviour are exhibited by a user in social networks: behaviour related to network graph structure, behaviour inferred from user created contents and finally silent behaviour recognizable from log data. Web mining approach is used to extract all of the above said behaviour pattern. There are three types of web mining approach applied for user behaviour analysis which are social network structure, content and web usage mining. Various hyperlinks available in the structure provide structural arrangements of the users. Web pages in the social network comprises of useful information. Log data in the server contains all the web usage, and these provide all the network access pattern of the users [4, 5].

Since the primary characteristics of a social network is highly dynamic in nature and behaviour is a qualitative term which represents a set of activities only, there is no straight forward mathematical formula which can directly represent a users behaviour. There are some prominent works in this regard in literature but everyone is capable of in certain aspects only. This article is organized into following different sections such as related work, nature of social network data, silent behaviour characterization and analysis, experimental results and analysis, limitations and future research directions and finally concluding remarks.

1.1 Related Work

As per literature available many people have tried to analyse social network user behaviour based on different concepts. Niranjana Kannan and Dr Elizabeth Shanthi in their research paper used classification and clustering for analysing navigation patterns of the users. They have used expectation maximization (EM) clustering over the log data for analysing the page visits and navigation behaviour in computer networks [6]. Reza Farahbakhsh et al., in their research work, characterized the user behaviour based on various mutual posts written in these three social network profiles [7]. They used the user-generated messages and profile data to characterize their common behaviour across the various social networks. Mei Li et al. have described various models of information diffusion behaviour in social networks in their research paper A Survey on Information Diffusion in Online Social Networks: Models and Methods [8]. They have cited various models including information diffusion model-based users structural arrangement in social networks. Wei Chen et al. in their research work, Mining hidden non-redundant causal relationships in online social networks, published in 2019, used a constrain-based data mining approach to find the users relationship behaviour in online social network [9]. Solomia Fedushko et al. in their research work, Modelling the Behavior Classification of Social News Aggregations Users, published in 2019, used fuzzy logic-based approach for user behaviour modelling [10].

There are some works where people have used the users profile data including regional and geographical information to analyse the user behaviour. Another type of behaviour people are interested in is anomaly user behaviour in social networks. Bimal Viswanath et al. had applied unsupervised anomaly detection techniques to separate anomalous user behaviour from original user behaviour in online social networks [11]. They used principal component analysis (PCA) technique for computing normal user behaviour and anomalous user behaviour. Analysis of behaviour characterization and representation in this paper is strictly confined to user's usage log-based measures irrespective of content and geographical area of the social network.

2 Nature of Social Network Data

By nature, social network data are unstructured which pose tremendous challenges to analysis to extract interesting patterns to satisfy various needs of researchers such as behaviour analysis, recommendation system, website design and organization. For characterization of user behaviour, mainly three types of data are available such as structural or graph data, content or textual data and log data [12]. All these social network data types are described below.

2.1 User Structure Data

This type of data is the website structure data designed to organize the content within the website. It includes intra-page structure of the content within a page represented through hyperlinks as tree structures. Here, pages are considered as nodes and links among the pages are vertices. Users' structural arrangement can be found out from the hyperlinks. This is called web structure mining. From these results, important web pages that a user visits in social network can be find out. Web mining technique can also be used for community discovery among the users having common interest in social network.

2.2 *User Content Data*

In social network, users exchange text messages among themselves for establishing relationship. These messages comprise of plaintext messages, videos and other web objects. Various formats include html/xml, video format, various scripts, some records from database, etc. Data mining can extract useful information from these web contents. Accordingly, these web content can be classified and clustered as per topic of interest. Traditional data mining can be applied for performing all the above tasks.

2.3 *Log Server Data*

Log data are collected in server using different methods. People use web crawler and application servers for capturing and storing log data. Every HTTP request generates one unique row of data entry in the logs of the server. Every log contains following fields: time and date, source IP address, name of the requested resource, web parameters used, request status, method involved in HTTP request, agent involved in user request, cookies present in client request, etc. Log data contain every click or user request, which is also known as web usage. While data mining technique is applied in web usage, it discovers the actual network access pattern. In this process, various data mining algorithms are applied. But pre-processing of log data is very challenging. Efficiently pre-processed data produce better results. From these data, users decision-making behaviour, browsing behaviour and popularity in the network can be computed. Social network log data have many formats according to server configurations. There are primarily three kinds of logs where data are stored. These are referrer log, agent log and server logs. The raw server log format is shown in Fig. 1.

3 Silent Behaviour Characterization and Analysis

Silent behaviour is observed neither in network structure data nor in basic profile data of users in the social network. These are captured from the server web log data for characterization of silent behaviour. Based on various silent social interactions

```
HC-192.168.164.121 19.8.2016 10:15 http://akz.imgfarm.com/images/webtooltab/chiclets/gmail.png
HC-192.168.164.121 19.8.2016 10:15 http://akz.imgfarm.com/images/webtooltab/chiclets/yahoo.png
HC-192.168.164.121 19.8.2016 10:15 http://akz.imgfarm.com/images/webtooltab/chiclets/facebook.png
HC-192.168.164.121 19.8.2016 10:15 http://akz.imgfarm.com/images/webtooltab/chiclets/youtube.png
HC-192.168.164.121 19.8.2016 10:15 http://akz.imgfarm.com/images/webtooltab/chiclets/amazon.png
HC-192.168.164.121 19.8.2016 10:15 http://akz.imgfarm.com/images/webtooltab/chiclets/twitter.png
HC-192.168.164.121 19.8.2016 10:15 http://akz.imgfarm.com/images/webtooltab/chiclets/wikipedia.png
HC-192.168.164.121 19.8.2016 10:15 http://akz.imgfarm.com/images/webtooltab/chiclets/pinterest.png
```

Fig. 1 A sample format of raw log data

such as visiting profiles, posting–deleting messages and login–logout, social network user's network access behaviour, navigation behaviour, etc., can be represented and analysed. Steps of this kind of behaviour representation and analysis methods are described in the following subsections.

3.1 Log Data Mining and Clustering-Based Network Browsing and Access Behaviour

This type of behaviour cannot be computed from profile data. All the silent activities leading to non-silent behaviour are not reflected in the profile data. Various silent activities such as navigation path and page access frequency are computed from the log data in the server. Similar activity patterns are detected from this log data. Silent behaviour is represented using clustering approach in weblog data. Some times users browse the pages doing no visible activities. In this case, no evidence remain with the profile data. Only traces of the browsing history are logged/recorded in server. From those logged data, browsing behaviour can be defined which are called silent behaviour in this study. So, silent behaviour includes the following activities to characterize and analyse user navigation patterns in online social networks (OSN)–frequency of page visits, time spent and sequence of activities performed, number of items visits in the site, etc. Clickstream-based analysis highlighted new insights on user behaviour in OSNs. All these activities are captured from the log data of social network users. From these log data, graph structure of all the activities can be computed using different centrality measures [13].

Log data are created by every click of the users in the network which are mined for useful information [14]. These log data are recorded in web server. Log data are plaintext in nature comprising of name of the user, user IP address, time record, referred URL and error codes if available, etc. Data recorded in log server are of different types namely transfer, agent, error and referrer log [15]. Most of the users clicked data are included in transfer and agent log. Again, error and referrer log are always not available and hence remain as optional. Log data contain the users' traversal data which contain IP address and other relevant information. There are three tasks involved in user behaviour characterization and analysis from log data. These are data collection and pre-processing, pattern discovery and pattern analysis. Here, in this work, pattern discovery and analysis have been given importance to represent user behaviour.

3.2 Session and User Analysis

User visits the social networks as per their convenience and interests which creates separate sessions per visits in a particular time frame. If a user logged in for long time doing nothing, then a standard time window is defined for identification of a session. Total users visited per session and total number of sessions created by a particular user are computed statistically for analysis of user behaviour. Data are segregated in different parts namely number of days, number of sessions, total visitors and domains. Various parameters such as user page accessed frequency, page view time, average path length, frequently used entry and exit gate and other related measures are computed using standard statistical tools. From this information, users are classified in some predetermined classes to analyse the user behaviour. Depending on the time window, user-activity records are segmented into various sessions where each session represents a single navigation to the social network. User-activity logs are partitioned and output a set of constructed sessions. Let us consider following social network user-activity log consisting of access time, machine IP address, home and referred URLs consisting of $\{P, Q, W, X, Y, Z\}$. User sessions are calculated using the following algorithmic steps:

Algorithm 1 Session identification algorithm

Input: Set of log data
Output: All the distinct and accepted sessions $S = \{s_1, s_2, \ldots, s_n\}$ as potential data clusters for every user $V = \{v_1, v_2, \ldots, v_n\}$
i. If a session time $t_i >= 300$ sec, where Session time $T_s = \{t_1, t_2, ..t_n\}$ and $i = 1, 2, ...n$ and number of accessed page is $>= 5$
ii. Classify these sessions as the potential data cluster
iii. Again if a session time < 300 sec and number of accessed page is $<= 5$
iv. Discard these sessions from the log data
v. end

Depending on the URL access time and visited URLs, user sessions are constructed as shown in Table 1.

From this user sessions, userwise segmented sessions are computed as shown in Tables 2 and 3.

3.3 Cluster Analysis and User Segmentation

User performs various activities in social network during an active session including visiting ones profile data, album, wall posts, etc. Since this study is limited to users silent behaviour, only silent activities are considered here for behaviour analysis. According to the set of activities performed by the users, clustering is performed for the segmentation of all the users. These clusters will aggregate all the users

Table 1 Structure of collected user sessions

Access time	IP address	Current URL	Referred URL
1:02	192.168.109.11	P	–
1:10	192.168.109.11	Q	P
1:20	192.168.109.11	W	P
1:26	192.168.109.11	Y	W
2:16	192.168.109.11	P	–
2:27	192.168.109.11	Z	W
2:31	192.168.109.11	Q	P
2:37	192.168.109.11	X	Q

Table 2 Separated user session-I

Access time	IP address	Current URL	Referred URL
1:02	192.168.109.11	P	–
1:10	192.168.109.11	Q	P
1:20	192.168.109.11	W	P
1:26	192.168.109.11	Y	W
2:27	192.168.109.11	Z	W

Table 3 Separated user session-II

Access time	IP address	Current URL	Referred URL
2:16	192.168.109.11	P	–
2:31	192.168.109.11	Q	P
2:37	192.168.109.11	X	Q

who show similar characteristics and browsing patterns [10]. This will infer user communities who represent identical interests in the user group and their activity behaviour in that particular social network. Let us consider, we have 10 number of users, $U = \{u1, u2, u10\}$ and 6 number of $URL = \{P, Q, W, X, Y, Z\}$. After clustering using any standard clustering algorithms like k-means based on URL accessed, we have following three clusters $CL = \{c1, c2, c3\}$ (Table 4).

3.4 Association and Correlation Analysis

Here, users' association with all the activities in all the transactions and sessions is analysed which is called association rule discovery and statistical correlation of users with their activities. User and activity association discovery is done based on apriori algorithms which find set of activities occurring frequently together in most of

Table 4 Separated users into different clusters

Cluster name	User	P	Q	W	X	Y	Z
c1	u2	0	0	1	1	0	0
c1	u5	0	0	1	1	0	0
c1	u8	0	0	1	1	0	0
c2	u1	1	1	0	0	0	1
c2	u4	1	1	0	0	0	1
c2	u7	1	1	0	0	0	1
c2	u10	1	1	0	0	0	1
c3	u3	1	0	0	1	1	0
c3	u6	1	0	0	1	1	0
c3	u9	1	0	0	1	1	0

Table 5 Activity transactions list

Transactions sequence
P, Q, X, Y
P, Q, Y, W, X
P, Q, Y, W
Q, Y, Q, P, W
X, P, Q, Y, W

the transactions. Associations which satisfy minimum such number (confidence) are then computed from the frequent activity sets. This information helps in doing users-activity behaviour analysis. For a given user session with a time window, algorithm considers all the frequent activity sets. It applies depth-first search over the activity set records represented as a graph. An activity recommendation value depends on the given confidence value of the respective user-activity combination rule. Following picture depicts a scenario of correlation and association rules (Table 5).

Based on this activity transaction sequence, all the activities are clustered based on their minimum support confidence as shown in the following Table 6.

3.5 *Sequential and Navigational User Behaviour Analysis*

Again another type of analysis is to find the sessionwise users sequential activity patterns. This computation will find the set of activity followed by another set of activities. From these sessionwise user-activity analysis, one can predict future visit patterns, overall behaviour analysis, etc., and also sequential pattern mining can be used to capture frequent navigational paths among user logs.

Table 6 Activity transaction clusters based on minimum support confidence

c1		c2		c3		c4	
Activity set	Support	Activity set	Support	Activity set	Support	Activity set	Support
P	5	P, Q	5	P, Q, W	4	P, Q, W, Y	4
Q	5	P, W	4	P, Q, Y	5		
W	4	P, Y	5	P, W, Y	4		
Y	5	Q, W	4	Q, W, Y	4		
		Q, Y	5				
		W, Y	4				

Table 7 Activity transactions list

Activity transactions	Frequency of occurrence
P, Q, Y	11
Q, X, Q, W	5
Q, W, Y	11
P, Q, Y, Z	7
P, X, Q	13
Q, X, Q, Z	9

3.6 *Behaviour Analysis Based on Sequence of Activities*

Sequential patterns of activities frequently occur in a sufficiently large proportion of transactions. Let us consider sequence set $S = s1, s2, \ldots, sn$ transaction set $T = t1, t2, \ldots, tn$. For a given transaction set T, the support (denoted by sup(S)) of a sequential pattern S in T is the fraction of transactions in T that contain S. The confidence of the rule GH, where G and H are (contiguous) sequential patterns, is defined as: $conf(GH) = sup(GH)/sup(G)$ Activity transactions and their corresponding frequencies are tabulated as follows (Table 7).

Like other activities, normal path followed by a user in navigation of a social network also leverage a particular behaviour if it is properly analysed over a long time period.

4 Experimental Results and Analysis

Social network user log data contain all the silent activities of the users. In this paper, we analyse users silent behaviour based on network access in terms of various sessions, login logout time duration, sequence of activity performed, etc. $C4.5$ method has been used for selection of potential sessions from the given log data. We considered all those sessions whose time durations are more than 5 min. We used facebook log data and nearly 89% accuracy has been achieved in filtration of potential sessions. Again based on the various sessions performed by the users, all the users have been clustered using k-means algorithm. We have seen that all these 3 prominent clusters generated include all the users. Some users are always active and perform many sessions in social network. These types of users are active users. Again some users login social network after a certain gap period and access the network silently. These types of users belong to second cluster. The last category users login social network very randomly after a long gap period. These types of users are less active in social networks. So, in this way, users may be categorized according to their silent behaviour in the network. Almost 90% accuracy has been achieved in clustering users using k-means algorithm.

While comparing our work with the behaviour modelling and classification [10], they defined the activeness of the users in the social network in terms of amount of content the user has created. They also considered some qualitative features such as creativeness, attractiveness, reactiveness and loyalty for measuring whether a user is active or not using fuzzy logic and classification. But in our work, we defined and measured the activeness of a user depending on its visit frequency and sessions in the social network. We used data mining algorithm in the log data for classification of the users as more active, medium and non-frequent social network users.

5 Limitations and Research Issues

There are many research directions in social computing especially in case of user behaviour computation from many aspects. As time passes, types of data collection, processing methods, storing criteria, etc., keep on changing with respect to social network platforms. Since social network is highly dynamic in nature, data analysis methods and algorithms with more efficiency including designing of efficient web crawler to extract and retrieve knowledge from the data sets have been always an open research problem forever. In this study, data collection methods, social network platform, etc., have not been considered while discussing analysis of silent user behaviour. Hence, social network platform and data collection method-oriented characterization of dynamic as well as static user behaviour can be a further studied in this direction. Again in our behaviour analysis approach, knowledge extraction from images and videos analysis have not been considered. Semantic analysis of data sets using natural language processing methods for behaviour characterization is also an important research area in social computing as it has many application in real life, which is out of scope of this article.

6 Concluding Remarks

This whole study of computable user behaviour analysis approach has been carefully designed in order to apply this approach in experiments to dig out as accurate and real as possible silent user behaviour in a given social network. Our effort has been put in place to accurately formulate some ways to leverage out user behaviour, although we know that behaviour is a qualitative term. Hence, accurate design of an algorithm is not sufficient, and a most proper qualitative analysis of the experimental results is utmost important to represent and discover real behaviour of an user in the social network platforms where experiments were performed.

Behaviour analysis and exact mathematical representation of behaviour are difficult as it is a qualitative analysis derived from computation. Proper clustering of users in the log data and computation of various metrics provide sufficient backgrounds for qualitative analysis of silent user behaviour analysis. But more concrete way of behaviour analysis requires accurate mathematical representation of various user activities in social networks.

References

1. Nieminen, J.: On the centrality in a graph. Scand. J. Psychol. **15**(1), 332–336 (1974)
2. Freeman, L.C.: Centrality in social networks: conceptual clarification. Soc. Networks **1**(3), 215–239 (1979)
3. Sabidussi, G.: The centrality index of a graph. Psychometrika **31**(4), 581–603 (1966)
4. Goel, N., Jha, C.K.: Analyzing users behaviour from web access logs using automated log analyzer tool. Int. J. Comput. Appl. **62**(2) (2013) ISSN: 0975-8887
5. Das, K., Sinha, S.K.: Centrality measure based approach for detection of malicious nodes in Twitter social network. Int. J. Eng. Technol. (To be published in June 2018)
6. Kannan, N., Shanti, E.: Classification and clustering of web log data to analyze user navigation patterns. J. Global Res. Comput. Sci. **1**(1), 36–40 (2010)
7. Farahbakhsh, R., Cuevas, A., Crespi, N.: Characterization of cross-posting activity for professional users across Facebook, Twitter and Google+. Soc. Network Anal. Mining **6**(1), 1–14 (2016)
8. Li, Mei, et al.: A survey on information diffusion in online social networks: models and methods. Information **118**(8), 1–21 (2017). https://doi.org/10.3390/info8040118
9. Chen, W., et al.: Mining hidden non-redundant causal relationships in online social networks. Neural Comput. Appl., pp. 1–11 (2019)
10. Fedushko, S., et al.: Modelling the behavior classification of social news aggregations users. https://dblp.uni-trier.de/db/journals/corr/corr1909.html (2019)
11. Viswanath, B., et al.: Towards detecting anomalous user behaviour in online social networks. In: 23rd Security Symposium (USENIX) Security 14, San Diego, CA, pp. 223–238, 2014, ISBN:978-1-931971-15-7
12. Das, K., Sinha, S.K.: Essential pre-processing tasks involved in data preparation for social network user behaviour analysis. In: Proceedings of the International Conference on Intelligent Sustainable Systems (ICISS 2017), IEEE Xplore, CFP17M19-ART, ISBN:978-1-5386-1959-9
13. Beauchamp, M.A.: An improved index of centrality. Behav. Sci. **10**(2), 161–163 (1965)
14. Bonacich, P., Lloyd, P.: Eigenvector-like measures of centrality for asymmetric relations. Soc. Networks **23**(3), 191–201 (2001)
15. https://en.wikipedia.org/wiki/Server_log (2019)